Issues in Indian
Agricultural Development

About the Book and Author

This work analyzes growth and structural change in Indian agriculture over the last three decades. In order to develop a global perspective, the Indian agricultural growth experience is introduced using parallels and contrasts with other parts of the Third World. The book is characterized by an empirical approach to the underlying economic data and a multi-disciplinary approach to the ramifications of agricultural growth. Considered among these are the transformation of the female labor force, population migrations and changes in human welfare. This book differs from the numerous others on Indian agriculture insofar as it takes a regional perspective, focusing on the causes and effects of inter-state variations.

Dr. M. Zarkovic is assistant professor of economics at St. Joseph's University, Philadelphia, Pennsylvania.

Issues in
Indian Agricultural
Development

M. Zarkovic

Routledge
Taylor & Francis Group

LONDON AND NEW YORK

First published 1987 by Westview Press, Inc.

Published 2018 by Routledge
52 Vanderbilt Avenue, New York, NY 10017
2 Park Square, Milton Park, Abingdon, Oxon OX14 4RN

Routledge is an imprint of the Taylor & Francis Group, an informa business

Library of Congress Catalog Card Number: 87-50898
ISBN 13: 978-0-367-01409-4 (hbk)

ISBN 13: 978-0-367-16396-9 (pbk)

Contents

Tables and Figures

Tables

Figures

Preface

My first research visit to India was concluded just prior to the lifting of emergency rule by Indira Gandhi, in the spring of 1977. It was a period characterized by anxieties about the prevailing political system and the staggering problems facing the economy. The evident need to confront issues such as poverty, hunger and illiteracy in a more aggressive and innovative fashion was apparent to government planners, academics and the general public. The frustrations of a nation groping for a new, better way to solve its basic problems are exemplified by a story popular in India at the time:

> India's development planners were very impressed with the progress made by Cuba and China following their "each-one-teach-one" literacy campaigns as well as the success of Israel's afforestation program based on the "each-one-plant-one" campaign. They were determined to come up with an original way of addressing India's problems, and thus they devised a campaign aimed at simultaneously solving their two major problems, population and hunger. The slogan of the campaign was to be "each-one-eat-one."

This anecdote reflects some of the discouragement that prevailed in India during the late 1970s which resulted from the achievements attained by other developing nations. India is often compared to China because these countries share demographic and geographical characteristics. This comparison clearly indicates the overall success of China's development strategy, especially with respect to growth rates, population control, various health indicators and literacy advances. Even countries with different conditions, such as the Newly Industrialized Countries of Asia (NICs: Taiwan, S. Korea, Singapore) have experienced economic growth spurts accompanied by poverty reductions uncommon in India's experience. Development scholars tend to stress the differences between countries which make each development effort unique. This uniqueness implies that the success of China and Cuba cannot be transferred to other political systems and that the replicability of the development model of the NICs is unlikely wherever the export orientation of the economy does not coincide with the demands of the predominantly western consumers. Thus, the few success stories of the developing countries are treated as inapplicable and irrelevant for India since *ceteris paribus* conditions never hold. The underlying attitude among development planners seems to be that each country, with its own set of problems, conditions and resources must find its own appropriate path of development.

This attitude was, in fact, a key element of India's approach to development over the past few decades. The resulting efforts to find the "Indian path" can be exemplified by its innovative experiments in the coexistence of the private and public sectors, central planning amidst decentralization as well as numerous sector-specific programs. In addition, the principal goals of governmental development efforts vacillated between food production, industrialization and poverty elimination. For example, the Second Five Year Plan was characterized by a zealous industrialization drive, whereas during the last few years of Indira Gandhi's rule, the issue of principal concern was the reduction of poverty. Recently, Rajiv Gandhi has receded from the direct attack on poverty, opting instead for the indirect approach associated with supply side economics. This entails the encouragement of modern production methods, free enterprise and greater tax incentives, with the aim of increasing production which in turn affects poverty and unemployment by way of the "trickle-down effect." These latest policies involve a new approach to old problems, and as such are embraced with the same hope that accompanied new approaches in the past. They represent yet another way in which India is groping for a development path appropriate for its particular conditions.

This book focuses on efforts to develop the agricultural sector. It contains a study of the economic growth experienced within agricultural India and the structural transformation of the economy that occurred as a consequence. The process of growth entails numerous ramifications that permeate the existence of rural inhabitants, such as changes in poverty, employment, migration, women's work, etc. These changes are assessed in this book from a multi-disciplinary perspective. The book is furthermore characterized by an empirical approach to the underlying economic data. In order to develop a global perspective, the Indian agricultural growth experience is studied using parallels and contrasts with other parts of the Third World.

Conventional books dealing with Indian agriculture do not address such a breadth of issues. The extensive literature on the subject most often deals with the technical aspects of economic and agricultural growth, or the class-polarizing effects of the green revolution. This study furthermore differs from most others in that it takes a regional approach to agricultural development, focusing on the causes and effects of *interstate* variations. This regional orientation is justified because of extensive decentralization among the states with respect to policy in areas such as land reform, social expenditure, rural industry, etc.

M. Zarkovic

Acknowledgments

My research on Indian agricultural development was conducted on two separate visits to India. During the first visit in 1977, I was affiliated with the Indian Statistical Institute (New Delhi), and I am most grateful to C.R.Rao for his introduction to India and his family's assistance during the course of my visit. It was during this visit that I also profited from the resources of the National Institute of Community Development, the International Crop Research Institute for Semi-Arid Technology and the Administrative Staff College, all in Hyderabad, as well as the Institute for Economic and Social Change in Bangalore. During the course of my research in Hyderabad, my understanding of Indian development problems was greatly enhanced by my discussions with Balwanth Reddy, to whom I remain indebted. During my second visit in the summer of 1984, I was affiliated with the Indian Agricultural Research Institute (New Delhi), for which I express my gratitude to A.S. Sirohi of the Economics Division, as well as H. Haque and D. Srivastava. This research was partially supported by a grant from St. Joseph's University. Lastly, in order to consult the latest data sources published by the Indian government, I utilized the resources of the London School of Economics in July 1985, while on a Summer Stipend from St. Joseph's University. Although the writing of this book was started in Philadephia, its completion was facilitated by a leave-of-absence from teaching responsibilities, for which I am deeply grateful to Tom McFadden and my colleagues at the Economics Department, St. Joseph's University.

I am most indebted to Jay Mandle for everything he has taught me during our discussions, which were a valuable source of stimulation for several topics in this book. Furthermore, I would like to thank David

Good and Sayre Schatz for their reading and their comments on various sections. In addition, I wish to express appreciation to Henri Barkey and Pat Conboy Kuzyk for helpful comments and engaging discussions of various conceptual and logistical problems arising throughout this book's creation. Finally, I want to thank Richard and Karla for their continuous support, their tolerance of my absences from home as well as dinnertime discussions about agriculture. There is no doubt that this book would never have materialized without Richard's patience, encouragement and assistance, for which I am so grateful.

M. Z.

1

Introduction

Outline of the Book

The chapters in this book address agricultural issues in the ten major agricultural states of India. By definition, these are the states in which the agricultural contribution to state income exceeded 50 percent in 1960. These states are: Andhra Pradesh (A.P.), Bihar, Haryana, Karnataka, Kerala, Madhya Pradesh (M.P.), Orissa, Punjab,[1] Tamil Nadu (T.N.) and Uttar Pradesh (U.P.). Although the importance of agriculture in state income ranges from a high 68.9 percent in Orissa to a low 52.0 percent in T.N., all states share problems and benefits of agricultural policies and innovations. However, it is essential not to overlook the diversity. The area covered by these states spans the entire Indian subcontinent and therefore geographic variations have lead to different agricultural practices and patterns of cultivation. Furthermore, the historical experiences of these states were by no means the same, resulting in dissimilar agrarian, economic and institutional structures. Lastly, due to their differing ethnic and religious composition, factors such as caste determinism in employment, taboos on work, migratory practices, etc., may play greater or lesser roles.

The omission of the four major states whose agricultural income falls below 50 percent, namely Gujarat, Maharastra, Rajasthan and West Bengal, is only problematic insofar as these highly industrial states are obviously linked to the agricultural states through the labor and product markets. These four states contain rural regions which clearly share characteristics of the agricultural states, but by virtue of their highly developed industrial base, a comparison on the *state* level with

1

predominantly agricultural states is inappropriate. It is only in chapter seven that the industrial states are included because the study of interstate migratory flows necessitates a discussion of industrial destinations of the migrant.

Not all agricultural states are observed in all the chapters. As a result of the uniqueness of Punjab, this state is sometimes studied in isolation. Indeed, Punjab lends itself to a study of agricultural development since it has the highest income per capita, the greatest agricultural production, and consequently the most pronounced evidence of the ramifications of growth.

A few words are in order pertaining to the terms in use throughout the book. Development scholars have often disagreed on the meaning of development and growth. Economic growth is a less provocative concept and usually implies a mere increase in income per capita (where income refers to state or national domestic product). It is the term "development" that warrants clarification. It is useful to divide existing definitions into two categories according to whether they are normative or positive in nature. In the normative perspective, economic growth *unaccompanied* by an improvement in the quality of life or standard of living does not constitute development. According to this view, development fails to occur if the lives of the majority have not been improved. It thus entails an expression of what "should" occur when a country grows, as well as an indication of the goals of development. This perspective is associated with development scholars such as Goulet and Blanchard (see chapter six). On the other hand, the positive approach to development does not incorporate considerations of the quality of life and limits itself to a description of the source and consequence of long term growth. According to this view, espoused by Kuznets (1966, 1973) and those in his tradition, development entails economic growth whose source is essentially technological in nature and whose outcome is a structural transformation of the economy. Kuznets does not distinguish between the concept of development and "modern economic growth,"[2] which he defines as obviously more than a mere increase in income per capita:

> ...[it] is a long term rise in the capacity to supply increasingly diverse economic goods to its population, this growing capacity based on advancing technology and the institutional and ideological adjustments that it demands.
>
> (Kuznets 1973: 165)

This view is shared by Friedman (1972: 84), who defines development as "an innovative process leading to the structural transformation of social systems" while growth is merely an expansion of the system without an alteration of the structure. This distinction between growth and development can be traced to the 1930s, at which time Schumpeter wrote:

> Nor will the mere growth of the economy, as shown by the growth of population and wealth, be designated as a process of development. For it calls forth no qualitatively new phenomena, but only processes of adaptation of the same kind as the changes in the natural data.
> (Schumpeter quoted in Gupta 1983: 68)

Therefore, economic development occurs when technology is the source of growth and when structural transformation of the economy ensues. This view of development has been adopted in this book. Considerations pertaining to welfare are raised separately in chapter six, but use of the word "development" by no means assumes it. In fact, the possible and sometimes probable discrepancy between development and improvement in welfare is the central theme of chapter six.

The book has been organized in the following fashion. Chapter one presents a review of post-Independence Indian agriculture and the economic growth experienced by the agricultural states. Chapter two assesses the degree of growth that was achieved and analyzes the structural transformation that occurred (or failed to occur), as well as the sources of that growth. The relative importance of technological change versus an increase in inputs of production is considered. Several factors (such as agricultural credit, the size and ownership of holdings, the product market, etc.) are introduced to aid in the understanding of why two states, Punjab and Haryana, adopted more innovative technology and experienced greater increases in income per capita than any of the remaining agricultural states.

Chapter three contains a study of the prevailing technology used in agriculture during the 1960s and 1970s, especially with respect to its effect on the labor force. The net effect on primary labor of different aspects of the green revolution technologies is empirically assessed in the high growth states of Punjab and Haryana, where these technologies were most widespread.

Another aspect of the relationship between technology and the labor force is the effect of the former on females residing and working in the rural areas. As a result of the recent emphasis on women's role in the development process, a new awareness of the positive contribution of females has emerged, as well as a realization of the detrimental ramifications the development process may have on females. Chapter four addresses these issues by assessing whether Indian females are made better or worse off by economic growth and the technological changes in agriculture. The study analyzes data pertaining to female employment and invokes cultural differences among the Sikh and Hindu farmers in its explanation of differing absorption and displacement rates among female workers.

A study of current conditions in India cannot overlook the question of Punjab and the Sikhs. It is suggested that the issue of a separate Punjabi country (Khalistan) transcends politics and religious orientation and is rooted in economics, specifically in the economic role of this region and its economic relations with the Indian Union. Punjab is the state with the highest level of income per capita, it is the greatest producer of foodgrains in South Asia and its tremendous agricultural growth has stimulated the expansion of industry. The relationship between agriculture and industry in this region is characterized by the importance of industries producing inputs for agricultural production, thus *enabling* agricultural growth. Given these characteristics of Punjabi agriculture and industry, demands for political independence entail dramatic economic consequences for both state and country. Chapter five addresses itself to the economy of Punjab, both with respect to agriculture and industry, and questions the economic viability of an independent Punjab.

The policy followed by the Indian government during the late 1970s and early 1980s reflected the general tendencies of the World Bank under the directorship of McNamara: there existed an emphasis on the *basic needs* of populations in developing countries, accompanied by an effort to address them directly with programs aimed at eradicating hunger, illiteracy and illness. The question posed in chapter six is whether the agricultural states of India experienced an improvement in the level of living (measured by various indicators of basic needs satisfaction) while they underwent economic growth. In other words, does economic growth result in an improvement in levels of education, health, etc.? Empirical evidence supports the contention that higher rates of growth *do not necessarily* imply greater improvements in the satisfaction of basic needs.

The explanation for this draws upon the role played by the public sector in each state and takes an historical view of differing state expenditures on social services.

In chapter seven, interstate migration is observed. The evidence from various countries indicates that migration occurs in response to economic stimuli and, in this way, a growing economy is able to meet its manpower requirements. It was questioned whether this relationship between economic growth and migration existed on a state level in India, and the relative importance of the "pull" exerted by industrial and agricultural employment is assessed.

The last example of the diversity of issues covered in this book is chapter eight, in which Marxist terminology is adopted in order to enter the ongoing debate on the *mode of production* in Indian agriculture. The three parts of the chapter include, first, empirical evidence for the existence of capitalist production, second, a theoretical link between capitalism, technology and economic growth (which points out the growth promoting tendencies of the capitalist mode of production), and third, an assessment of the implications for human welfare of a capitalist mode of production.

Finally, chapter nine contains some conclusions pertaining to the topics above, as well as some general comments on the resulting interrelationships.

These chapters contain topics relevant in the study of the source, the process and the consequences of agricultural growth. These are not unrelated topics, but are instead linked by Kuznets's theory of modern economic growth. Briefly, according to Kuznets, growth tends to have the following characteristics: first, its primary source is technological change, second, it results in the structural transformation of income and labor, and third, the environment best suited for this growth is the one loosely described as capitalism. This book covers exactly these topics as they relate to agriculture: technology and the source of growth is addressed in chapters two and three, the structural transformation in chapters two and five (with respect to income), and with respect to the labor force, in chapters three (all labor), four (specifically female labor) and seven (labor migration). Lastly, the environment most suited for growth is discussed in chapter eight (capitalist mode of production) and six (growth and human welfare in various economic systems). Thus the Kuznetsian theoretical framework, as applied to Indian agriculture, defines the underlying cohesion of the topics under study.

Overview of Indian Agriculture

Background on India

India is a highly populated, relatively poor country, whose economic structure reveals the dominance of agriculture and where economic planning occurs within the context of a mixed economy. Although this description fails to mention many aspects of this highly diverse and rich country, it does include some pertinent points which warrant elaboration.

The population of India reached 749 million in 1984, and by conservative estimates is expected to reach one billion by the end of the century.[3] Although the population growth rate has decreased in the recent years to just under 2 percent, the effect of decades of growth between 2.5 and 2.3 percent resulted in a large dependency ratio.[4] The Indian population is heterogeneous with respect to its ethnic, linguistic, religious and cultural composition. Eighty percent of this population resides in the rural areas, and mostly under conditions of staggering poverty: it is estimated that 40 to 50 percent survive below the poverty line.[5] Although this poverty line refers to nutritional requirements (income to enable the purchase of a predetermined number of calories per day), other characteristics of poverty further corroborate this point. For example, the adult literacy rate for the whole country is merely 36 percent (1981), while the distribution indicates the highest rates among urban males and the lowest among rural females. Furthermore, poverty cannot be discussed in isolation of employment: 53.2 percent of the males and 20.8 percent of the females are in the labor force, amounting to 37.6 percent of the total population (247 million people in 1981).[6] Of these, 8 percent are currently unemployed, amounting to approximately 20 million. In addition, the degree of underemployment has been conservatively estimated at 18 million people. Lastly, poverty can be simply described by income per capita, which in 1983 was US$260, thereby placing India among the low income developing countries according to the World Bank's classification. The average annual growth in the preceding two decades was 3.9 percent per annum, low in comparison with the 5.3 percent average growth of all countries in the low income category.[7] In fact, international comparisons have shown that India's growth is surpassed by countries that share characteristics such as population, resources and colonial experience (i.e., Brazil, Taiwan, South Korea, China and Thailand).

The Indian economy is characterized by a mixture of centralization amidst decentralization. The Constitution clearly lays out the economic spheres under the control of the center and the states. The Planning Commission, in its Five Year Plans, identifies the country's objectives and methods for their attainment, sets targets for production and social indicators, and offers guidelines for the states to follow. In addition, the Indian economy is described as a "mixed economy" insofar as it consists of private and public ownership of productive assets. These two sectors are viewed as complementary, since individual effort and private initiative are considered important stimuli, whereas government involvement is considered necessary in industries and services which may fail to develop on their own. The relative importance and freedom allotted to the private and public sectors has oscillated throughout the post-Independence era (through an extensive system of government controls such as liscencing and price setting), but the economic activities encompassed by each have changed little: small scale industry, agriculture, trade, construction and most services fall into the private domain, whereas the public sector consists of basic and heavy industries (including steel, coal and fertilizers), as well as banking and insurance.

The agricultural sector is of importance in the Indian economy both with respect to its share of national income, as well as in its role as employer. In 1979-80, 34.4 percent of the national income was derived from agricultural activities (compared to 25.7 from industries and 38.9 from services), indicating a dramatic change over the proportions observed in 1950-51: 56.1, 17.3 and 26.6 percent respectively (V.K.R.V. Rao 1983: 34). With respect to employment of Indian labor by sector, over two thirds of the labor force derives its livelihood from agriculture and allied activities.

It is clear that agricultural growth rates are crucial to the Indian economy, as is the nature of the agricultural production process and its effect on employment. Agriculture's importance is further corroborated by evidence that demand from the rural consumers is a principal source of stimulation for the industrial production of consumer goods. It follows that agricultural income is a critical determinant of the pace of manufacturing expansion. On the supply side, agricultural production determines the extent of food availability, as well as inputs into non-agricultural production, such as raw materials. Lastly, it should be noted that agriculture contributes to the production of agro-industrial goods,

which presently account for 33 percent of industrial production
(Balasubramanyam 1984: 48).

Given this obvious importance of the agricultural sector in the Indian
economy, it is understandable that studies on the subject have proliferated
in the recent decades. The literature ranges in political orientation from the
"left" to the "right", addresses itself to causes of agricultural problems and
offers suggestions for solutions. Scholars such as Rudra, Omvedt,
Patnaik, and Bagchi, tend to view the persistence of rural poverty, *despite*
agricultural growth, as the result of capitalist relations both nationally
(based on the exploitation of one class by another), as well as globally
(based upon exploitation of one country by another). Other scholars such
as Khurso, Hannumantha Rao, Raj Krishna, Chaudhary, as well as
Bauer, Bhagwati, Chakravarti and Schultz, have focused their
explanations of rural stagnation on structural factors such as land
distribution, or institutional factors such as the nature of center-state
relations, the credit system and the infrastructure. However, regardless of
their orientation, these scholars share a condemnation of the post-
Independence development pattern with respect to both growth and
poverty. This development process and its achievements are briefly
described below, with special reference to the agricultural sector.

Evolution of Indian Agriculture

The review of the historical evolution of Indian agriculture in the
modern period should begin with an observation of the impact of British
colonial policy on the Indian development experience. Although the
overall impact has been judged in the literature as mixed, the stimulative
effect of the introduction of a railway system, the construction of ports,
and the incorporation of India into the world economy through trade is not
without merit. Specifically with respect to agriculture, the principal
objective of the Raj consisted of the transformation of the Punjab region
into the breadbasket of South Asia by means of heavy investment into
irrigation facilities. Despite this effort, the last three decades of the Raj
were characterized by stagnating agricultural production. Blyn (1966)
estimates that grain production increased a mere 0.03 percent during the
three decades preceding Independence, while rice output declined at an
average rate of 0.09 percent.[8] In addition, the per capita availability of
foodgrains between 1911 and 1941 declined by 26 percent. These data
should be viewed against the total growth picture provided by Heston

(1983), according to which the rate of growth of the total economy during this period was 1 percent, while no growth occurred in income per capita.[9]

Under these conditions of overall stagnation and inadequate agricultural production, it is understandable that Independence in 1947 was accompanied by enthusiasm for the transformation of rural areas. The community development projects, which proliferated throughout the country, reflected an effort to engage the local population in the development effort.[10] Partially as a result of limited budget and partially to fully harness the population's enthusiasm at the time of Independence, the community development projects grandiosely aimed at totally transforming the rural areas through self-help. According to Neale (1985: 677), they entailed "giving the old idea of development from below new form and content". In this spirit, the First Five Year Plan (1951 to 1956) allocated 32 percent of its budget outlay to agriculture, community development, and irrigation. The latter was viewed as a supporting element of community development insofar as it was a prerequisite for increased production, and its introduction to new regions was expected to alter the stagnating production patterns while agricultural technology remained unchanged.

At the inception of the Second Five Year plan (1956 to 1961), the results of the community development strategy were not yet clear, although some preliminary evidence indicated disappointment was warranted. Criticisms of the project were based upon a realization that the importance attached to a project of such magnitude was too meagre, that an intersectoral approach might have stimulated linkages with agriculture, and that human resources and enthusiasm were not sufficient for rural transformation. The Second Five Year Plan nonetheless contained elements of the "do-it-yourself" philosophy. Although the total budget for the Second Five Year Plan was greatly increased, agriculture's share did not grow by the same proportion: indeed, the percent share of agriculture decreased dramatically, although it did increase in absolute terms.[11] The controversy over the relative importance of agriculture and industry emerged at this time since it was during the Second Five Year Plan that industries were granted primary importance. This resulted from the application of Mahalanobis's (1953) ideas of rapid economic development through intensive investment in basic and heavy industries. The central government did attempt to rectify the neglect of agriculture by stimulating rural production through institutional changes. Among these was the eradication of exploitation by village moneylenders, landlords, and various other rural intermediaries through land reforms and zamindari abolition acts.[12]

The success of the total agricultural strategy of this period varied from region to region, but essentially was assessed both by scholars as well as government investigators as dismal. Hindsight has indicated the difficulties inherent in the eradication of exploitative relationships in remote rural areas where they have existed for centuries and where there is no recourse to the law formulated in New Delhi. The success of community development projects was also varied throughout the country, but it became evident that the hurdles to be overcome in order to effectively implement this policy are enormous, especially given the infrastructure characterized by illiteracy, generations of bribery and corruption, and the concentration of power and wealth in the hands of the landlords. Meanwhile, agricultural production picked up from its pre-Independence low, but failed to reach the projected levels: agricultural production grew by 2.7 percent (average annual compounded growth) during 1950-1960, whereas the total economic growth failed to surpass 3.8 percent during this period (Rao 1983: 32). Despite this increase over the pre-Independence period, there is evidence that the overall rural situation did not improve and that agriculture's full potential was not exploited. It became clear that a change in priorities and method for improving agricultural sector performance was warranted.

This change in Indian agricultural policy came about as a result of research on farmer economic performance which demolished the myth of the "tradition-bound peasant" (Jones 1960, T.W. Schultz 1964, Krishna 1967).[13] It was suggested that Third World farmers are rational, economic agents that will respond to price incentives, change production techniques and adjust output levels *when it is rational to do so* in order to maximize their profits and minimize their costs. An application of this view of the farmer to concrete policy entailed the creation of a stimulating environment in which the profit maximizing tendencies of the farmers can be unleashed. This implied that the supply of technology and appropriate price incentives on part of the government would result in farmer performance characterized by increased production, increased employment, decreased poverty, and increased income. However, such a policy was too all-encompassing to be aimed at *all* farmers in *all* regions simultaneously (this reflected Hirshman's (1958) defense of unbalanced growth: if it were possible to develop all regions at the same rate at the same time, there would be no problem of development). Thus, the "growth-center" approach was chosen, and twelve districts throughout the country were isolated to become the pioneering recipients of the Intensive Agricultural Development Programme (IADP). These regions were already predisposed to growth insofar as their infrastructure was best developed and, according to cynics, were regions in least need of development efforts. Therefore, as part of the Third Five Year Plan (1961 to 1966), the agricultural priorities of the planners focused on the introduction of sufficiently attractive production technologies to stimulate the response of the rational farmer. Increased agricultural production became of crucial importance, given the unpredictable weather conditions and the subsequent food shortages. Although attributing rational economic behavior to illiterate farmers is common today, two decades ago it represented a major shift in the formulation of rural economic solutions.

During the early 1960s there was another change in the agricultural sector which was to have a profound effect on production, the food crisis, and income distribution in many developing countries. The high yielding varieties (HYV) of wheat and rice seeds, developed in Mexico and the Philippines, were being adapted to the Indian setting. The introduction and proliferation of these seeds became a priority of agriculture in the Fourth (1969 to 1974) and Fifth Five Year Plans (1974 to 1979). Although the resulting "green revolution" consisted of new varieties of many crops, it is only rice and wheat that are relevant in the Indian context insofar as their potential, as well as their success, was the greatest.[14]

The green revolution in all its complexity has been discussed with extreme vehemence and passion, perhaps because of the promises it held. Essentially, the literature can be divided into two parts: one part addresses the economic implications of the proliferation of technological inputs, especially its ramifications on production and growth, and the other part focuses on the social aspects of innovative techniques. From the point of view of growth and output, evidence indicates that in all-India, for all green revolution crops, the agricultural growth rates during the period 1967 to 1981 were lower than the average during 1949-50 to 1964-65 (Balasubramanyam 1984: 83). However, this general statement of stagnation conceals regional and crop differences: during the first ten years of HYV applications, wheat was the only crop that achieved high rates of growth, leading to Srinivasan's conclusion that the green revolution was no more than a "wheat revolution."[15] Output of wheat increased from 10.9 million tons in 1961 to 23.8 million tons in 1971

(Government of India 1983: 228). It was not until 1975 that rice experienced a similar take-off. Success in achievement of high rates of growth in output were concentrated in various pockets of the country, essentially in wheat producing Punjab, Haryana and western Uttar Pradesh (U.P.).[16] These regions were responsible for approximately 70 percent of total wheat production during the late 1960s. Therefore, it was not about either rice or wheat *in isolation*, nor Punjab and Haryana, that scholars such as Hannumantha Rao[17] referred to when they claimed that the green revolution was irrelevant in its impact on growth, but rather to the general trend in agricultural production which obviously did not warrant the term "revolution" associated with it.

Despite the possibility of dividing the green revolution in India into the wheat and rice revolutions, the social problems resulting from both are essentially identical. Primarily, the green revolution has been condemned because it further exacerbates the already unequal distribution of wealth and productive assets existing in most developing countries.[18] With increased use of the technology, the gap between the small and large farmer tends to grow, as well as between the landlord and his tenant. Simultaneously, the gap between geographical regions increases. Despite this, there is no consensus that the net social effect was negative because of the enhanced food availability which has contributed significantly to the improvement of the rural existence. Thus, the characterization of the green revolution as a success or failure depends upon which of the above aspects is deemed primary.

As a result of disappointment associated with the unrealized growth rates in agriculture, as well as the increased polarization of income groups, the central government introduced the Integrated Rural Development Programs (in 1978-79). The purpose of these was to redress problems arising from a policy of intensive application of resources to certain self-contained regions while neglecting others where social and economic problems multiplied.[19] The disparity and agricultural stagnation was viewed with sufficient concern by the Janata government that 40 percent of public investment was allocated to the agricultural sector in the late 1970s. Agricultural policy was aimed at target groups estimated to be adversely affected by the green revolution (earning less than 3500 Rs. per year for family of five) under the following subprograms: Drought Prone Area Program, Integrated Tribal Development Projects, Small Farmer Development Agency, and the Agency for Marginal Farmers and Agricultural Laborers, among others. The Sixth Five Year Plan (1980 to

1985) addressed itself in earnest to the rural unemployment problem by the introduction of the National Rural Employment Programme in 1980 to replace the Food for Work Program and to complement the Employment Guarantee Schemes (which attempted to stimulate wage employment and income generating self-employment).

Despite four decades of planning and observation of the agricultural sector and the rural population, the Indian government has yet to come to grips with rural poverty, unemployment, and in some regions, the low rates of agricultural growth. Although the production of agricultural goods is essentially in private hands (some cooperative production exists but it is by no means widespread),[20] the privately accumulated income is not channeled into the improvement of infrastructure such as transportation facilities or storage, which therefore remains the domain of government efforts. Institutional constraints such as credit availability and marketing arrangements are known to be deficient but the magnitude of the problem tends to paralyze planners. Perhaps the new economic policy proposed by Rajiv Gandhi in the Seventh Five Year Plan (1985 to 1990) will result in some positive changes. However, it is already clear that the agricultural sector is not envisioned as a priority sector, and is treated as a part of a complex system of economic interrelationships. The benefits to agriculture are expected to trickle in from the liberalization of business and industry.[21] Although agricultural development remains a goal, the agrarian lobby has been dissatisfied with the new plan insofar as it is considered to not only neglect the rural population but in fact to harm it. Specifically, there has been a decrease in all subsidies to agriculture except in the case of fertilizers. In addition, the tax relief that is the cornerstone of the new plan, *fails to reach the farmers* since it is limited to direct taxes which farmers do not pay, while indirect taxes such as excise taxes have been increased. According to Rubin (1985: 953), the only aspect of the Seventh Plan that the agricultural lobby can be pleased with is the introduction of crop insurance. This evidence does not bode well for rural India, which is still expected to provide food, employment, and consumer demand for the rest of the economy.

An Assessment of Indian Agriculture

Four goals have consistently underlied Indian agricultural strategy over the past decades, although their relative importance has been in flux. These are: the achievement of a targeted level of growth, the elimination of unemployment, the eradication of poverty and the attainment of self-sufficiency in food. The first three goals are discussed at length throughout the book, and it will suffice to say here that their successes and failures depend upon the spatial and temporal dimensions of the observations. While there is no doubt that the total number of workers in agriculture has increased during the course of 1961 to 1981, the population expansion was so great that the rural participation rate rarely increased. One result of this mounting pressure on agriculture to provide livelihood for increasing numbers of people is the persistence of poverty. The overwhelming evidence indicates that, with respect to the elimination of rural unemployment and poverty, socio-political factors have prevented the realization of efforts to aid the weaker sections. Too often, the wealthier farmers have reaped benefits of governmental efforts. The structural distribution of land and assets is in their favor, which tends to go hand in hand with political and bureaucratic power. It has thus been argued repeatedly that in the absence of a major change in the power structure, as well as property relations, little progress can be made in the achievement of these goals (see, for example, Dantwala 1979).

The situation with respect to growth rates and food self-sufficiency is perhaps more hopeful. In the case of the former, the period 1973 through 1984 has been characterized by 2.3 percent average annual growth in agricultural production, whereas the production of foodgrains averaged 2.7 percent, slightly ahead of population growth. In comparison to India's past record, this represents success. Success is also evident upon comparison with the growth rates in Korea, Argentina, Greece or Yugoslavia. However, many countries have achieved greater growth rates. The obvious example that comes to mind is China, where the recent introduction of the "responsibility system" into agricultural production resulted in growth rates of over 7 percent per annum. Although it is highly unlikely that these rates can be sustained over a long period of time, it is a fact that the Chinese rural areas experienced a transformation that is, as yet, missing in India.

The growth in foodgrain production of 2.7 percent can be translated into real values: output jumped from 89 million tons in 1963-64 to 108 million in 1970-71, mainly as a result of wheat production. After a brief stagnation, rice yields took off, resulting in a peak production of foodgrains of 131 million tons in 1975-76. Again, another stagnation period set in and persisted until 1982-83, when both rice and wheat experienced growth spurts resulting in 152 million tons in 1983-84 and an expected 160 million in 1986.[22]   Clearly, the growth in foodgrain production was not smooth. It was characterized by fluctuations, in particular during 1965-1967 and 1972-73, resulting mainly from erratic weather conditions. During the first drought period, India was compelled to import 24 million tons of wheat from the U.S. Although this represented a peak, each year until 1982 (with the exception of 1972) witnessed the importation of foodgrains. Currently, the situation has so drastically reversed itself that not only is there no need to import food, but India is exporting a few million tons of wheat to the Soviet Union, as well as rice and wheat in small quantities to Bangladesh, Nepal, Maldives, etc.[23]   In addition, India is in a position to donate food (for example, 100,000 tons of wheat was given to Ethiopia in 1985).   However, the existence of foodgrain exports does not imply that the food problem for the rural Indian peasants has been solved. Indeed, not only has the increased supply *not* resulted in a price decrease, but the government floor price for wheat has increased in order to keep farmers in production. Furthermore, the rice producing states are substituting wheat for rice in consumption, thus decreasing the demand for wheat from the wheat producing regions of north India. As a result, an increasing portion of wheat output is currently siphoned into buffer stocks which have reached 35 million tons in 1986.

It is therefore incorrect to view the Indian agricultural situation with either extreme gloom or exhilaration. Progress has unequivocally been achieved in the attainment of some goals, but progress is often accompanied by a new set of problems which must be addressed against the backdrop of ever increasing rural populations whose demands on the economy represent a drain *rarely* offset by their positive contribution. The Indian agricultural development experience has shown that "one must run just in order to remain in the same place." Currently, serious efforts are made to reverse this trend, efforts primarily focusing on the creation and introduction of innovative technology (biotechnology) which, like the green revolution in its time, is capable of dramatically transforming the rural scene.[24]   Technological developments in food production are moving at a pace far surpassing those of the past, as seeds resistant to the vagaries of monsoons and diseases are emerging on the market, together with low cost fertilizers, and various inputs appropriate to local conditions. Currently, biotechnology holds the greatest hope for increasing yields by bypassing most of the traditional constraints in crop production (such as marginal land, weather induced stresses, costly irrigation schemes). Indeed, India is foremost among the developing countries in the promotion of research on this new agricultural technology, recognizing that it is breakthroughs of this nature that hold the key to improved rural existence and consequently to growth of the whole economy.

TABLE 1.1

General and economic characteristics of India

characteristic	year	value
life expectancy	1983	(total) 55 years
		(male) 56 years
		(female) 54 years
population	1984	749 million
urban population	1983	176 million
		24% of total
GNP (US$)	1980	162 billion
GNP per capita (US$)	1980	240
rate of inflation	1973-1983	7.7%
gross domestic investment (ave. annual growth)	1973-1983	4.2%
current account balance (US$)	1983	-2780 million
net inflow of external capital-public (US$)	1983	1995 million
external public debt (US$)	1983	7940 million
		11.2% of GNP
growth of trade (average annual)	1973-1983	(exports) 4.9%
		(imports) 2.8%

Source: various tables in the Annex of World Bank, (1985), *World Development Report* Washington.

TABLE 1.2

Population, area and population density in Indian agricultural states, 1981

State	Population (x 1000)	Area (sq km x 1000)	Pop. Density (per sq km)
A.P.	53,550	275.1	195
Bihar	69,915	173.9	402
Haryana	12,923	44.2	292
Karnataka	37,136	191.8	194
Kerala	25,454	38.9	655
M.P.	52,179	443.5	118
Orissa	26,370	155.7	169
Punjab	16,789	50.4	333
T.N.	48,408	130.1	372
U.P.	110,862	294.4	377
India	685,185	3287.3	216

Source: Government of India, Ministry of Information and Broadcasting, (1983) *India,* New Delhi, Table 1.2.

Notes

1. The principal territorial change in post-Independence India that affects the data in this study is the division of Erstwhile Punjab into Punjab, Haryana and Himachal Pradesh. This occurred in 1966 in response to linguistic and religious demands made by the Sikhs and Hindus of these regions. As a result, it was sometimes necessary to adjust pre-1966 data for the boundary changes, or failing that, to adopt observations made in the first year following the division, i.e., 1967.

2. Mandle, a scholar in the Kuznets tradition, is instrumental in clarifying the distinction made by Kuznets between growth (referring to a mere increase in income per capita) and "modern economic growth" (Mandle 1980).

3. Ambannavar's projection is considered in the medium range, coinciding with the World Bank's estimate of 994 million (J.P. Ambannavar, (1975), *Second India Studies: Population,* Bombay: Macmillan Co. of India).

4. It is estimated that in 1981, 43 percent of the Indian population is outside of the working years of 15 through 64, and thus dependent on

others for survival (World Bank, (1985), *World Development Report*, Washington, p.214).

5. There is much controversy about how the poverty line should be determined. Variety exists in the estimates of population residing below the poverty line, depending upon the assumptions that are made concerning the minimum nutritional level, the prices of goods needed to reach that level, and the income that is necessary to pay those prices. The principal studies on the Indian poverty line range from Sukhatme's (1978) optimistic view (15 percent of rural inhabitants are below the line) to Bardhan's (1974) contention that 54 percent of the rural inhabitants live in absolute poverty, to Minhas's (1974) estimate of 37 percent and lastly, to the study by Dandekar and Rath (1971), indicating 40 percent of the rural population experience absolute poverty. These are all reviewed in Balasubramanyam (1984).

6. It must be noted that the definition of labor is very generous and encompasses casual and marginal labor in addition to full time workers.

7. The statistics in the preceding paragraphs are derived from various tables in the World Bank, (1985), *World Development Report*, Washington, and Government of India, (1983) *India* New Delhi.

8. G. Blyn, (1966), *Agricultural Trends in India 1891-1947: Output, Availability and Productivity*, Philadelphia: University of Pennsylvania Press.

9. Alan Heston, (1983), "National Income" in Kharma Kumar (ed.), *The Cambridge Economic History of India*, London: Cambridge University Press.

10. Holdcroft (1978) discusses the movement in support of community development programs in over sixty countries of Asia, Africa and Latin America during the 1960s. He attributes its popularity to its non-revolutionary approach to agrarian development.

11. The First Five Year Plan (FYP I) allocated 3570 million Rs. to agriculture, which was raised to 5680 million Rs. in the Second Five Year Plan (FYP II). In terms of percentages of the total outlay, agriculture received 15.1 in the FYP I and only 11.8 in the FYP II. Dantwala (1979: 6) points out correctly that the criticism of agriculture's neglect in the FYP II can just as easily be applied to industries during the FYP I, when it received a mere 6.3 percent of the budget (and rose to 14.4 during the FYP II).

12. There exists an extensive literature on land reforms in India, both from the point of view of their desirability as well as an assessment

of their success (see Bandyopadhyay 1986, Joshi 1975, Uppal 1983, among others). However, it should be noted that there is a consensus among scholars as to the dismal effect of the reforms that have to date been enacted as well as the pressing need for further change in the agrarian structure.

13. "Farmers are everywhere capable of producing the right things in the right place in the right amounts and at low costs in terms of resources if they receive the proper economic signals" (from T.W. Schultz, (1964), *Transforming Traditional Agriculture*, New Haven: Yale University Press, p.29). A similar view is presented in the following: W.O. Jones, (1960), "Economic Man in Africa" in *Food Research Institute Studies* 1, and R. Krishna, (1967), "Agricultural Price Policy and Economic Development" in H. Southworth and B. Johnston (eds.), *Agricultural Development and Economic Growth*, Ithaca: Cornell University Press.

14. Jowar, bajra and maize seeds have also been improved during the green revolution, but their success and use is limited to 3.8, 4.5 and 1.6 million hectares, respectively. This is low in comparison to the area under rice (19.7) or wheat (16.8) in 1981-82 (Government of India 1983: 236).

15. This issue is raised in T.N. Srinivasan's "The Green Revolution or the Wheat Revolution?", *Indian Statistical Institute* Discussion Paper #66, New Delhi, 1971, as well as in "Trends in Agriculture in India, 1949-50 to 1977-78," *Economic and Political Weekly*, August 1979.

16. One repercussion of this dramatic difference between the productivity and output of wheat and rice has been the further strengthening of the division of states according to their deficit or surplus status. Although this distinction between states has existed since at least 1957, the green revolution contributed to the change in the size of the deficit or surplus (see P. Spitz, (1983), Food Systems and Society in India: A Draft Interim Report (vol. 1), Geneva :UNRISD, pp. 163-168).

17. It is suggested that agricultural growth would have occurred anyway because population growth would have exerted upward pressure on agricultural prices, which in turn would have stimulated an increase in supply, and would have induced the government to invest in the agricultural infrastructure. See C.H. Hanumantha Rao, (1975), *Technological Change and Distribution of Gains in Indian Agriculture*, New Delhi: Macmillan Co. of India, Ltd.

18. Although the literature on the subject is vast, Griffin's study (1974, *Political Economy of Agrarian Change: An Essay on the Green Revolution,* Cambridge: Harvard University Press) is excellent for an overview of the general topic, as is Hayami's (1981, "Induced Innovation, Green Revolution and Income Distribution: Comment" in *Economic Development and Cultural Change,* 30/1). Frankel (1971) and Prahaladachar (1983) address the issue in the particular context of India.

19. These programs reflect a general tendency throughout the developing countries to increase production and increase social services. However, by the 1980, many governments and donors had retreated from IRDPs and started giving greater emphasis to agricultural production. Holdcroft (1978) compares their worldwide rise and fall to that of the community development program.

20. By 1981, there were 300,000 cooperatives in operation in India, mostly in the rural areas, with an estimated membership of 106.2 million people. However, these numbers include all cooperative efforts, such as credit societies, marketing and processing cooperatives, as well as societies for the sharing of agricultural implements and irrigation facilities. Production cooperatives are very rare.

21. See "Rajiv Gandhi's New India" in *South,* October 1985.

22. The statistics in this paragraph are taken from the article, "When more means less" in *South,* March 1986.

23. Statistics pertaining to foodgrain exports in this paragraph are taken from Government of India 1983: 233.

24. A comprehensive review of the potential of biotechnology, as well as its possible social ramifications, is presented in a study by Buttel et al. (1985).

2

State Economic Growth: Sources and Ramifications

This chapter addresses the causes and effects of state economic growth. The analysis of the growth process is based on the pioneering research of scholars such as Kuznets (1966, 1973), Denison (1974) and Kendrick (1961),[1] who provided guidelines for monitoring economic activity. Their research focused on the determinants of economic growth, the measures of the growth process, and the ramifications of economic growth throughout the economy. Their findings provide a generalization of the Western growth experience, one that offers insights into the process but does not necessarily serve as a standard for judging performance in different situations. Its applicability to the Indian case is examined in this chapter. In the first section, the structural transformation of the state economies is described. In the second part, an analysis is conducted into the relative importance of technological change and mere input increases in the determination of the source of economic growth.

Economic Growth and Redistribution of Economic Activity

The research of Kuznets (1966) indicates that a long term increase in income per capita tends to result in fundamental changes in the structure of an economic system. Specifically, the importance of the three sectors (agriculture, manufacturing and services)[2] changes such that the proportion of national income originating in agriculture decreases, whereas the share of the manufacturing sector increases dramatically. Income from services increases slightly or stays constant. There exist two reasons for these changes, associated with income elasticity of demand

and the nature of technological change. With increased income per capita, consumers tend to demand less agricultural products and more manufactured goods and services whose income elasticity of demand is greater than that of agricultural goods. It follows that producers respond to the increased demand by increasing production, hence that sector's share of state income increases. With respect to the technological reason for structural transformation of sectoral income, it has been found that innovation tends to be most extensive in the manufacturing sector, followed by agriculture and lastly by services. Since greater efficiency of production results in greater generation of income, it follows that the sectoral share of manufacturing in total income tends to increase most rapidly.

Thus, changes in demand and technological innovation combine to determine the structural transformation of the sectoral share of income in the economy. Income data from the agricultural states of India are presented in order to determine whether the process of structural transformation occurred over the period 1961 through 1981.

Evidence from India

The national income accounting methods developed by Kuznets were adopted by Indian scholars to estimate changes in economic variables. The works of Mukherjee (1969) and V.K.R.V. Rao (1983) stand out as the leading studies covering different periods in the Indian post-Independence growth experience. On the basis of their research it can be assessed that during the past three decades, the Indian growth rate averaged 3.5 percent per annum and was characterized by wide fluctuations, coinciding with weather conditions as well as the wars with China and Pakistan. Furthermore, it should be noted that the growth rate of the last fifteen years was in fact lower than that achieved during the first fifteen years of planning (1950-51 to 1964-65: 3.7 percent per annum, compared to 3.0 percent per annum achieved during 1966-67 to 1979-80, excluding the drought years of 1965 through 1967). The total growth experience of all-India is impressive with respect to its past (average annual rate of 1 percent during the fifty years preceding Independence) but pales before the rates achieved by countries such as Sri Lanka, Kenya, Pakistan, Indonesia, where the growth rate ranged from 4.5 to 6 percent (World Bank 1982).

Mukherjee's study made use of India's principal source of income estimates, the Central Statistical Organization (CSO), whereas V.K.R.V. Rao relied more heavily on the Reserve Bank of India (RBI) statistics.

The CSO, a branch of the Planning Commission, publishes data on growth rates and the industrial origin of income. It uses the nine-item classification of industries,[3] and combines the product and income approach to measuring domestic income. The RBI income statistics include a breakdown by state, using data collected by the State Statistical Organizations. It is the RBI data that were used throughout this chapter, despite their limitations due to state differences in methods of collection and compilation of data, as well as differing deflators and base years.[4] The state domestic product in the ten agricultural states is presented below.

(1) Total Income Levels

A review of state income, as measured by absolute and per capita levels of state domestic product (SDP), indicates the fluctuations in income that occurred during the course of the past two decades, as well as the relative position of the ten agricultural states (Table 2.1). Although the values in constant rupees take into account the inflation rate, the time series are not strictly comparable because base years sometimes differ. For this reason, the values for 1970-71 are presented two times, once with a 1960-61 base, and again with a 1970-71 base.

A review of the absolute levels of constant SDP in 1960-61 shows that U.P. and T.N. were the leading states (18,430 and 11,120 million Rs. respectively) followed by Bihar and A.P. (9,930 and 9,830 million Rs. respectively). Haryana, Kerala, Orissa and Punjab had relatively low levels of total income throughout the study, irrespective of the base year.

However, the ranking of the agricultural states according to SDP per capita is not the same as that described above: Punjab, T.N. and Haryana had the highest per capita income at the beginning of the study (347, 334, and 327 Rs. respectively), all above the Indian average of 307 Rs. By 1970-71, it became clear that Punjab and Haryana are dramatically different from other agricultural states (with income per capita of 471 and 441 respectively, compared to 356 Rs. in all-India). This difference became even more pronounced during the 1970s using 1970-71 as a base year. Bihar and Orissa experienced the lowest income per capita at the onset of the study, and retained this position irrespective of the base year.

TABLE 2.1

State domestic product, in constant Rs.

absolute income (in 10 million Rs.)

State	(a) 1960-61	(b) 1965-66	(c) 1970-71	(d) 1974-75	(e) 1970-71	(f) 1980-81
A.P.	983	1083	1335	1550	2523	3447
Bihar	993	1118	1123#	na	2245	3095
Haryana	245	275	437	463	868	1351
Karnat.*	556	637	905	1063	1985	2337
Kerala	432	488	626	705	1255	1562
M.P.	832	786	1075	1172	1991	2552
Orissa	374	465	574	576	1037	1383
Punjab	383	460	632	719	1436	2280
T.N.	1112	1174	1427	1358	2371	2954
U.P.	1843	2019	2359	2392	4256	5687
all-India	13335	15234	19282	20281	32744	44460

per capita income

State	(a) 1960-61	(b) 1965-66	(c) 1970-71	(d) 1974-75	(e) 1970-71	(f) 1980-81
A.P.	275	276	310	330	586	649
Bihar	215	221	207#	na	402	447
Haryana	327	319	441	421	877	1061
Karnat.	238	243	306	327	685	637
Kerala	259	261	298	304	654	619
M.P.	260	217	261	259	484	493
Orissa	217	239	265	245	478	529
Punjab	347	378	471	496	1070	1380
T.N.	334	317	349	305	581	615
U.P.	252	252	269	254	486	519
all-India	307	314	356	345	633	697

note: columns a-d: base year 1960-61, columns e-f: base year 1970-71.
* base year 1956-7
# refers to 1969-70
Source: columns a-d: *Reserve Bank of India Bulletin*, April 1978, statement 2.
columns e-f: Planning Commission unpublished data.

(2) Income Growth

The growth rates of per capita income (Table 2.2) further corroborate the remarkable performance of Haryana and Punjab during the 1960s, in particular during 1965-66 through 1969-70: 6.8 and 5.7 percent compounded average annual growth, compared to the all-India value of 0.5. The growth rates dropped dramatically during the 1970s (according to either base year), although relative to other states and to all-India, Haryana and Punjab remained the highest growth states (1.9 and 2.6 average annual growth, compared to the all-India value of 0.97).

The growth of the total SDP in Punjab and Haryana followed the same pattern as that of per capita income both with respect to changes over time and rank relative to other states. The average annual growth rates in both Punjab and Haryana slowed down in 1970s from their growth during the 1960s: from 5.1 to 4.7 and from 6.0 to 4.5 respectively. Upon comparison with the remaining agricultural states, it is clear that the experience of Punjab and Haryana is unmatched.

(3) Income by Sector

The agricultural states of India have undergone changes with respect to the proportion of income derived from each sector. Table 2.3 shows that all states experienced a decrease in the percent of income derived from agriculture, although in T.N. this decrease was most remarkable (from 52.0 to 41.8 percent). The manufacturing sector increased in all states, although the largest increase was observed in T.N. and Bihar (from 17.6 to 24.2 and from 10.6 to 18.0, respectively, in percentages). Services increased in importance in all states except in Bihar, where its contribution to state income did not change during the 1960s.

Therefore, the sectoral performance of states indicates that growth *did* result in the structural transformation of the economy as was claimed by Kuznets to be a characteristic of modern economic growth.

TABLE 2.2

Growth rates of state domestic product (compounded yearly average) in constant Rs.

absolute income

State	(a) 1960-1 to 1970-1	(b) 1965-6 to 1969-70	(c) 1970-1 to 1980-1
A.P.	3.10	.24	3.17
Bihar	na	1.97	3.26
Haryana	5.96	11.47	4.52
Karnat.	4.99	3.89	1.65
Kerala	3.78	4.42	2.79
M.P.	2.60	2.66	2.51
Orissa	4.38	1.57	2.92
Punjab	5.14	3.91	4.73
T.N.	2.53	2.38	2.23
U.P.	2.50	1.83	2.94
all India	3.76	2.68	3.11

per capita income

State	(a) 1960-1 to 1970-1	(b) 1965-6 to 1969-70	(c) 1970-1 to 1980-1
A.P.	1.20	1.59	1.03
Bihar	na	-.88	1.07
Haryana	3.04	6.75	1.92
Karnat.	2.54	2.81	-.72
Kerala	1.41	2.05	-.54
M.P.	0.04	0.16	0.18
Orissa	2.02	0.43	1.02
Punjab	3.10	5.74	2.58
T.N.	0.44	0.27	0.57
U.P.	0.65	0.08	0.64
all-India	1.49	0.46	0.97

note: columns a,b: base year 1960-1, column c: base year 1970-1
Source: columns b,c: *Reserve Bank of India Bulletin*, 1978, statement 7, columns a,d: derived from Table 2.1.

TABLE 2.3

Income by sector, in constant Rs.

base year 1960-61

State	1960-61			1970-71		
	I	II	III	I	II	III
A.P.	58.8	12.8	28.4	53.5	16.2	30.3
Bihar	58.0	10.6	31.4	51.2[a]	18.0[a]	30.8[a]
Haryana	62.9	16.1	21.0	59.4	18.4	22.2
Karnat.	62.1	12.0	25.9	50.9	17.8	31.3
Kerala	56.0	15.2	28.8	50.6	17.0	32.4
M.P.	63.4	14.9	21.7	55.8	19.5	24.7
Orissa	63.2	12.4	24.4	60.3	13.5	26.1
Punjab	57.0	16.5	26.5	54.7	17.5	27.8
T.N.	52.0	17.6	30.4	41.8	24.2	34.0
U.P.	60.2	11.1	28.7	55.3	14.7	30.0
all-India	52.2	19.1	28.7	49.8	20.2	30.0

base year 1970-71

State	1970-71			1978-79		
	I	II	III	I	II	III
A.P.	53.5	16.0	30.5	45.7	20.2	34.1
Bihar	62.8	16.8	20.4	56.7	21.3	22.0
Hary.[b]	59.4	18.4	22.2	52.7	21.1	26.2
Karnat.	59.6	18.6	21.8	57.0	19.0	24.0
Kerala	54.7	16.4	28.9	46.3	21.1	32.6
M.P.	61.6	15.2	23.2	55.5	17.8	26.7
Orissa	68.9	12.0	19.1	67.8	11.5	20.7
Punjab	60.3	13.3	26.4	57.4	15.0	27.6
T.N.	40.9	25.6	33.5	38.2	26.9	34.9
U.P.	60.3	14.9	24.8	57.3	16.4	26.3
all-India	na	na	na	35.4[c]	25.7[c]	38.9[c]

note: [a] refers to 1969-70

[b] base year 1960-61 for all Haryana values

[c] refers to 1979-80

Source: various tables in *Reserve Bank of India Bulletin*, 1978 and 1981, and
c: Rao 1983: 34.

The Source of Economic Growth

Economic growth can occur as a result of technological change[5] or an increase in the inputs used in the production process. Most often, these inputs entail labor, capital, land, and raw materials. Technology, often referred to as "the coefficient of our ignorance" because of the difficulty with which it is identified and measured, entails either the introduction of new (or improved) inputs into the production function, or a change in the way in which the old inputs are combined. The effect of technological innovation, whatever its form, tends to be more efficient production, implying greater output at same cost or same output at lower cost than prior to the innovation.

Substantial empirical evidence exists to support the contention that most economic growth experienced by the industrialized world can be attributed to this technological innovation. Davis's (1964: 33-60) study of the U.S. economy indicates that over the course of 1900 to 1960, 38.5 percent of growth was due to labor changes, 2.5 percent to increases in land use, 19.2 percent to the use of more capital (assuming homogeneous inputs), and the remaining 42.5 percent could not be accounted for by any increase in inputs but has been attributed to innovative techniques in production.[6] Similar results were attained by Solow (1957) for the U.S. economy during 1909 to 1949, by Denison (1974) for the period 1929 to 1969 and Kendrick (1961, 1973) for the periods 1889 to1957 and 1948 to 1969.[7] Studies of other countries and time periods have also pointed to the importance of technology (Denison 1967, UNECE 1964).[8] Given this evidence, the traditional focus on capital accumulation seems inappropriate and insufficient without an emphasis on the technological input. This is not to say that innovation is always positive: in fact, there is evidence that the addition to the technological base does not always result in increased income and product. Saturation or inappropriate innovation is common, especially in developing countries where the adoption of western technology often occurs without adaptation to local conditions, resulted in the concept and application of "intermediate technology."[9] Despite arguments against rapid technological change, there is currently a concensus that "intermediate technology" is simply incapable of producing the desired growth rates and that capital accumulation is not the key to long term economic growth. Instead, the greatest potential for development lies in the productivity advances associated with technological innovation.

<u>Data from Agricultural India</u>

Accounting for economic growth by factor is usually accomplished through estimation of factor elasticities. This is done by observing compensation received by each input and attributing the share of growth which cannot be assigned in this way to the technical coefficient in the aggregate production function. Even in the industrialized countries, where data collection and tabulation is sophisticated and reliable, productivity accounting is nonetheless a difficult task wrought with approximations. However, in the case of India, these methods are even more difficult to apply given the lack of sufficient and reliable data to support a breakdown of productivity. In fact, as late as 1983, V.K.R.V. Rao used crude methods of attributing growth to factors of production by merely observing the change in the value of output per unit of input for all factors, and comparing it to the increase in income. The principal shortcoming of this method is that it does not allow for complementarity between inputs, yet it is the best that could have been accomplished given the data limitations. A similar analysis is performed in this study. The relative contributions of land, labor and capital are assessed while management is omitted for lack of approximate indicators and appropriate data. The percentage increase in each of these factors has been compared to the increase in income per capita over the same period. This simple method, although obviously lacking in sophistication, yields an approximation of the source of growth in agricultural income.

For the production of agricultural goods, several inputs are used irrespective of the output produced: land, labor, implements, seeds, and water management. The indicator of land adopted in this study is total area cropped, which is preferable to net area insofar as it incorporates multiple cropping on a single plot of land. Another important factor of production in Indian agriculture is labor, and the increase in the number of primary sector workers is an indication of the increased use of this input. However, the embodied improvements in both land and labor tend not to be noticed. In other words, if labor becomes healthier and more skilled, or if land is used in conjunction with other inputs so its productivity increases, these changes fail to show up in the percent change in factor use. The issue of capital is even more complex and leaves room for further speculation. Capital used in agricultural production takes on many forms, including HYV seeds, mechanical inputs such as tractors and threshers, as well as irrigation, fertilizers and pesticides. However, the *introduction* of these to a farm represents technological change. Although seeds are a capital input that has been in use for centuries, the HYV represent an innovation over the old seeds and are therefore distinguished from mere capital inputs. As a result, capital will not be measured *per se*, whereas tractors, irrigation, fertilizers and HYV seeds are treated as technological inputs, implying that land and labor are most clearly defined as inputs, whereas capital is distinguished from technology only with difficulty.[10]

Table 2.4 shows the percent change in income over the course of the 1960s as well as the percent change in the various inputs discussed above.[11] Clearly, Punjab and Haryana, followed by U.P. and Orissa, experienced the greatest percent change in their income per capita. With respect to the increase in the labor input that might have contributed to the growth in these states, only Punjab's agriculture absorbed labor during the 1960s, while in Haryana, U.P. and Orissa, the agricultural labor force actually decreased. Kerala and A.P., the only other states where the absolute labor force increased, experienced relatively mediocre increases in income per capita. During the 1970s however, all states except Kerala absorbed workers in their agriculture, and the two high growth states did not distinguish themselves from the low growth states with respect to labor absorption. With respect to changes in the land input, during the 1960s Kerala and M.P. experienced 14.5 and 15.3 percent increases in area cropped, whereas the two highest growth states, Punjab and Haryana, increased their land use by a mere 6.3 and 4.2 percent respectively. During the 1970s, this percentage increased in almost all states, partially as a result of the incorporation of area sown more than once into the category of area cropped, thus accounting for multiple cropping associated with the green revolution. According to this expanded concept of land use, Punjab and Haryana had the greatest increase in area cropped.

With respect to the technology/capital inputs, Punjab emerges as the primary adopter of mechanized inputs, irrigation and fertilizers during the 1960s. Although the increase in the use of tractors during 1966 to 1972 was similar in Punjab, Haryana and Kerala, the level of tractor ownership in 1972 as well as 1977 was remarkably higher in Punjab. This state also had the highest irrigated area as a percent of total area cropped. The increase in irrigated land during the 1960s was the greatest in M.P. (55.0 percent), although a mere 10.9 percent of its cropped area was irrigated in 1980-81. With respect to fertilizer use, Punjab leads the agricultural states (47.3 kg per ha), followed by T.N. (33.6 kg per ha in 1974-75). Haryana, another high growth state, used remarkably little fertilizer (15 kg per ha).

TABLE 2.4

Percent change in income and various sources of growth

(i) percent change in income per capita

	1960-61 to 1970-71	1970-71 to 1980-81
A.P.	12.7	10.8
Bihar	-3.7	11.2
Haryana	34.7	21.0
Karnat.	28.6	-7.0
Kerala	15.1	-5.4
M.P.	0.4	1.9
Orissa	22.1	10.7
Punjab	35.7	29.0
T.N.	4.5	5.9
U.P.	6.7	6.8

(ii) percent change in primary labor force[a]

	1961 to 1971	1971 to 1981
A.P.	16.9	10.7
Bihar	-4.0	10.8
Haryana	-13.7	25.5
Karnat.	-8.5	22.8
Kerala	56.0	-19.3
M.P.	-10.2	22.2
Orissa	-4.9	17.2
Punjab	26.3	15.2
T.N.	-1.7	21.4
U.P.	-2.4	13.2

(iii) area cropped

	% change 1960-61 to 1973-74	% change 1970-71 to 1980-81	area (100,000 ha.) 1980-81
A.P.	7.9	6.7	123
Bihar	4.3	32.8	112
Haryana	4.2*	53.9	55
Karnat.	0.1	4.5	107
Kerala	14.5	34.9	29
M.P.	15.3	16.4	214
Orissa	6.5	43.5	88
Punjab	6.3*	67.9	68
T.N.	3.0	6.6	65
U.P.	-0.1	40.6	246

(continued)

Table 2.4 (cont.)

(iv) tractor ownership

	% change 1966 to 1972	% change 1972 to 1977	tractors(x 100) 1977
A.P.	116.4	76.2	111
Bihar	162.6	87.5	105
Haryana	279.4	65.2	304
Karnat.	148.4	45.6	83
Kerala	258.9	13.3	17
M.P.	99.0	202.0	151
Orissa	169.9	-22.2	14
Punjab	298.3	57.3	667
T.N.	64.7	20.4	65
U.P.	172.2	158.3	713

(v) irrigated area

	% change 1960-61 to 1970-71	% change 1970-71 to 1980-81	irrigated area (%)[b] 1980-81
A.P.	9.6	8.6	28.2
Bihar	10.5	29.6	26.5
Haryana	na	51.6	39.1
Karnat.	33.3	19.0	12.8
Kerala	18.8	na	8.3
M.P.	55.0	63.0	10.9
Orissa	5.1	18.3	13.9
Punjab	na	19.3	53.2
T.N.	1.8	2.5	39.7
U.P.	34.4	38.6	38.5

(vi) fertilizer consumption [c]

	% change 1960-61 to 1974-75	% change 1974-75 to 1982-83	kg per ha 1982-83
A.P.	183.7	3.7	25.3
Bihar	916.7	-50.0	12.2
Haryana	na	297.7	59.5
Karnat.	251.6	105.5	22.4
Kerala	142.6	20.2	27.4
M.P.	433.3	12.5	5.4
Orissa	610.0	-19.7	5.7
Punjab	1291.1	256.2	168.5
T.N.	205.5	3.0	34.6
U.P.	709.1	39.9	49.6

(continued)

Table 2.4 (cont.)

(vii) area under HYV

	% change 1975-76 to 1980-81	area (x 1000 ha) 1980-81	% area[d] 1980-81
A.P.	24.4	3686	34.3
Bihar	55.0	3650	43.9
Haryana	42.2	2165	60.1
Karnat.	16.7	1863	18.8
Kerala	38.5	360	16.5
M.P.	58.6	3615	19.3
Orissa	118.1	1254	20.5
Punjab	39.5	4035	96.3
T.N.	34.0	2795	52.2
U.P.	26.8	8030	46.6

note: a: the percent change in the primary sector labor force incorporates the main and marginal workers of the 1961 census and the main workers of the 1971 enumeration
b: irrigated area as a percent of area cropped
c: fertilizer consumption refers to total chemical fertilizers, including nitrogen, phosphate and sulfuric complexes
d: area under HYV as a percent of net area sown
* 1965-74
** 1968-69 to 1973-74
Source: (i) derived from Table 2.1,
(ii) derived from various tables in *Indian Census* 1961, 1971, 1981,
(iii) derived from *Statistical Abstract of India* 1974, 1984,
(iv) reproduced from Table 2.10,
(v) derived from *Statistical Abstract of India*, 1964, 1974, 1984
(vi) Statistics Division, Fertilizer Association of India, *Fertilizer Statistics*, New Delhi; 1970-71 (Tables 6.15 and 6.17), 1974-75 (Tables 6.01 and 6.03),and 1982-83 (Table 8.04)
(vii) ibid. 1982-83 (Table 10.02).

According to the data on cultivation of HYV seeds by state, it is clear that by 1980-81, most of the area sown in Punjab was under HYV (96.3 percent), followed by Haryana (60.1 percent). However, this refers to all crops, and a breakdown by crop indicates that by 1980-81, 98 percent of the land used for wheat cultivation in Punjab and Haryana was under HYV seeds (Johar and Singh 1983: 14), a value higher than that for rice, or for any crop in the other agricultural states. Data by HYV crop and by state pertaining to the early period of the green revolution is sparse. Naidu (1975: Table 9.31) claims that in 1970-71, 75.9 percent of the irrigated area in Punjab was covered with HYV seeds, compared to 70.9 in Haryana and 44.0 in U.P.

The link between green revolution technology and economic growth has been drawn by many scholars who observed the increase in output of wheat, rice and other crops following the introduction of the various inputs described above. In the specific case of Punjabi wheat production, output increased from 1.9 to 5.6 million tons during the years 1965 through 1972 (Gill 1983: 2). With respect to the whole country, the Indian government attributes the increase in foodgrain production (from 72.4 to 133.0 million tons during 1965 through 1982) to the proliferation of green revolution technologies. The case of wheat is even more remarkable: 10.4 million tons were produced in 1965-66, compared with 36.3 in 1978-79 (Government of India 1983: 236). These numbers indicate that, despite some degree of technological saturation and unsuccessful application, production *has* increased. This increased product translates into increased income, and the change in this income per population is, in effect, a measure of economic growth.

Adoption of Technology

It seems clear from mere observation of statistics that Punjab, and to a lesser degree Haryana, adopted more green revolution technologies than the remaining agricultural states. Why did this occur? Extensive research has been conducted on farmer adoption of innovative agricultural technology (surveyed in the excellent article by Feder et al., 1985), and some of its salient elements are introduced below. Various observers of Indian agriculture have argued that state differences in adoption rates are due to the difference in the success of some HYV seeds. Desai (1971) and Sen (1974) hypothesized that wheat seeds respond better than either rice or jowar to supporting inputs (such as irrigation, fertilizer, etc.) which are an integral part of the package of technologies. Desai further claims that Punjab and Haryana adapted the imported wheat strain to local Indian conditions, a process facilitated by the geographical concentration of wheat producing states and consequently similar ecological conditions. Agricultural research facilities, already existing in these areas, were quick to experiment with the new crops and to provide extension services. This argument, based on crop choice, furthermore contends that the initial success of the crop provided a strong demonstration effect that induced farmers to adopt HYV of wheat. Rice, on the other hand, was unsuccessful for analogous reasons: it was not adapted to local conditions and tastes first because there existed few facilities for experimentation and second because the rice producing areas are spread out throughout the subcontinent, thereby exacerbating experimental efforts due to the lack of uniformity of conditions and tastes (Mellor 1976: 59). Therefore, while wheat yields were taking-off, rice production was stagnating. Rice farmers had less incentive to adopt the new technology than wheat producers who could count on good results, and consequently rice growing states experienced lower rates of economic growth than wheat producing states.

To test the validity of this argument, the principal crop in each state was identified by magnitude of output and accordingly, Punjab and Haryana are classified as wheat producing, whereas the remaining agricultural states cultivate mostly rice. An exception to this classification is U.P., where the western half of the state is predominantly wheat producing, whereas the eastern regions produce mostly rice. More area is cultivated with rice, but the output of wheat is higher, thus U.P. will be treated as wheat producing with the understanding that aggregating over the whole state tends dilute the effect of prevailing trends in any of the separate regions.[12]

During the decade of the 1960s, the three wheat states grew by an average of 25.7 percent, compared to 11.4 percent achieved by the rice states.[13] A comparison of the levels of technology adoption shows that wheat states adopted relatively more technology during the 1960s than the rice states: in 1970-71, 84 percent of cropped land in wheat states was under HYV seeds, compared to only 32 percent in the rice states, 50 percent of the area in wheat states was irrigated, compared to 22 percent in rice states, and fertilizer consumption was 25 (kg per ha) in wheat states, compared to 16 (kg per ha) in the rice producing states (Zarkovic 1984: 58). Clearly, the wheat states adopted more technology. Furthermore, *this technology resulted in high productivity of all crops, not only wheat*, as evident from the following two comparisons:

1. The average output of rice was observed in both rice and wheat producing states (Table 2.5). In the latter, rice is a secondary crop, given secondary consideration with respect to land quality and water resources. Under these conditions, the average output of rice in 1961-62 in the rice states was 11.44 (qu per ha), compared to 9.7 in the wheat states. Yet, by 1975-76, these values increased to 14.3 and 18.5 respectively. It is evident that the wheat producing states were more productive during this period *even in the production of their minor crop*.

TABLE 2.5

Average output of rice and wheat (in quintals per hectare) in rice and
wheat producing states

	Rice		Wheat	
	Rice States	Wheat States	Rice States	Wheat states
1961-62	11.4	9.7	5.6*	7.4
1966-67	10.6	9.4	4.1*	8.1
1971-72	13.7	15.5	12.2**	11.6
1976-76	14.4	18.5	10.4**	11.7
1977-78	14.5	23.5	10.5**	12.5

note: * refers only to six out out seven rice states
** refers only to five out of seven rice states. Data unavailable for the
omitted states.
Source: derived from various *State Statistical Abstracts* and Zarkovic
1984: 59.

2. The average output of rice and wheat was compared. By 1977-
78, rice states produced an average of 14.5 (qu per ha) of rice, compared
to 23.5 in the wheat states. With respect to wheat, the rice states produced
10.5 (qu per ha), compared to 12.5 in the wheat states. It is clear that
during the green revolution, *the production of rice in wheat producing
states exceeded the production of wheat in rice producing states.*

Therefore, those scholars who claim that high yields of wheat are
essentially the result of a better strain of crop may be in error. Wheat
performance was poor in non-wheat producing states, whereas rice did
well in non-rice producing states, suggesting that the critical element in
agricultural growth is not crop choice, but that the answer lies elsewhere.
Given the complexity of Third World farming systems, it is necessary to
turn to some microeconomic aspects of agricultural development. Among
these is a study of rural institutions. Several institutional characteristics
were observed in the agricultural states. These are: human capital, access
to capital for innovation, prices in the crop market, ownership of land and
size of cultivated holdings. These rural institutions undergo changes in the
course of development, and as a result alter incentives and access to
factors of production, and may explain why wheat states adopted more
technology than rice states.

(1) Human Capital

One of the conditions for technological innovation to proliferate in a region is that farmers be aware of the benefits it may bring. This awareness comes about with increased education, which is a prerequisite for understanding potential benefits from change and a determinant of the ability to comprehend new techniques and thus adopt new practices. Farmers must, according to Schultz (1981: 25), have "...the ability to perceive, interpret and respond to new events..." Schultz's 1964 work on adoption of technology inspired many scholars to empirically test the relationship between education and health on the one hand and innovation on the other. The literature on the subject suggests that farmers with better education are earlier and more efficient adopters of modern technologies (Evenson 1974, Villaume 1977).[14] This was supported in the case of HYV in Kenya (Gerhartt 1975), and in the case of chemical inputs in Thailand (Jamison and Lau 1982).[15] Welch (1978) suggested that the contribution by the farmer to the returns from production can be attributed to ability and allocative skills, both of which improve as experience and health improves.[16] Ram (1976) hypothesized that the value of education increases with technological change, and found that returns to education are higher in progressive regions than in the more backward areas (both in India and the USA).[17] Indeed, according to Welsh (1978), education has no impact on productivity in regions with traditional agricultural practices but is related positively to education in regions that are undergoing modernization. Other research has focused on the effect of education on adjustment to changes in prices, and both Huffman (1977) and Petzel (1976) show that more educated farmers adjust land and fertilizer use to changes in prices of inputs and outputs more readily.[18]

In the Indian context, both Chaduri (1968) and Sidhu (1976) found that formal schooling was relevant in the determination of worker ability: the former suggested that differences in education explain variation in cropping among regions, whereas the latter found education to affect yield and gross sales. In addition, Rosenzweig (1978) claims that the probability of adoption of high yielding grain in Punjab is positively related to education.[19]

Thus, the evidence from global studies as well as those particular to India, indicates the importance of education (formal, informal or extension) in farming. Literacy is one of the indicators of education. It can be stated that literacy, although by no means a prerequisite for cash crop farming, entails a certain awareness of society, and may be viewed as an indicator of receptiveness to green revolution technology. Given the performance of wheat producing states with respect to the proliferation of technological inputs, it might be expected that these states are characterized by higher literacy rates than the rice producing states. However, this is not the case. According to the data presented in Table 6.5 (in chapter six), Punjab, Haryana and U.P. in fact ranked third, fifth and ninth out of the total ten agricultural states in 1981, with literacy rates of 40.9, 36.1 and 27.2 percent respectively. Instead, rice producing Kerala achieved 70.4 percent literacy, followed by T.N. (46.8 percent). This seems to imply that *on the state level*, literacy is not a significant determinant of agricultural innovation.

(2) Credit Constraints

Development scholars of the 1950s viewed the financial constraint as a major impediment to development. This view persisted with respect to agriculture through the 1960s as a result of the heavy reliance on purchased inputs associated with the green revolution. Since agricultural investments have their source in either accumulated savings or capital markets, if access to these differs among farmers and regions, it follows that the rate of innovation also differs. There is evidence across developing countries that rural savings have been increasing. Specifically in India, Krishna and Chaudhari (1982) found that the marginal propensity of farmers to save rural income has increased from 1.33 to 11.79 percent during the course of 1951 to 1973. However, savings are rarely sufficient for innovation because they vary positively with size of the farming operation which tends to be small scale in the Third World. Consequently, access to financial markets is critical for most farmers.[20]

In an effort to determine the relationship between credit and technological innovation on a state level in India, loans advanced by agricultural cooperative societies were observed. These are the main source of credit in the rural areas, with the possible exception of the village moneylender. From Table 2.6, it is clear that Punjab had unequivocally the highest value of loans advanced per capita (in 1974-75, 58,000 Rs), followed by Haryana (41,000 Rs.). U.P., on the other hand, is among the states with the lowest value of loans advanced (9,000 Rs). Since this highly populous state is only partially wheat producing, and there exists no breakdown of credit by crop, the evidence with respect to wheat farmers in U.P. is inconclusive. However, it can be said with certainty that Punjab and Haryana, in the midst of the green revolution, had greater credit advanced per capita than the rice states, thereby reflecting the financial ability of their farmers to adopt new technology.

TABLE 2.6

Loans (per capita) advanced by primary agricultural credit and agricultural societies, in current Rs (x1000)

	% change 1960-61	% change 1974-75
A.P.	5.3	7.2
Bihar	0.4	2.9
Haryana	na	41.0
Karnat.	6.0	21.2
Kerala	3.0	20.8
M.P.	5.5	15.7
Orissa	1.5	8.8
Punjab	5.5	57.7
T.N.	7.2	21.3
U.P.	4.2	8.5

Source: derived from various tables in various volumes of Fertilizer Association of India, Statistics Division, *Fertilizer Statistics*, New Delhi.

(3) Price Incentives

An important determinant of the supply of wheat and rice is the price of both output and inputs. The rational farmer will produce a particular crop or adopt a particular innovative practice on the basis of a cost/benefit analysis. Numerous supply models have been empirically tested and the resulting evidence indicates a positive relationship between price of output and marketable surplus. Reca (1983) found that a 10 percent increase in prices of agricultural output results in a 2 to 4 percent increase in output, *ceteris paribus*.[21] Sidhu's (1979) research indicates that marketed surplus response to price changes is positive in the case of food, and with specific reference to wheat, Evanson (1977) found that a 10 percent increase in prices induced a 3.2 percent increase in production.[22]

There exists ample evidence from India that farmers respond to price incentives. Krishna (1967) estimated the elasticities of substitution of acreage with respect to prices and found that Indian farmers shift into the production of new crops when the price is right. T.N. Krishnan (1977) also found a positive response of marketed surplus of food with respect to price among Indian farmers. Furthermore, price has a greater positive price among Indian farmers. Furthermore, price has a greater positive influence on production of rice and wheat than on other food crops: on the district level, Punjab, A.P. and T.N. showed the most consistent pattern of positive estimates.[23] Given the close relationship between an increase in supply and the adoption of the HYV package of technologies, especially in wheat and rice regions, prices in the product market are an indicator of the incentives influencing the farmer.

Do the findings for individual farmers apply when aggregated over states? Did the level and fluctuations of foodgrain prices stimulate the production of wheat more than rice? Was the expected profit from wheat production greater than in the case of rice and for that reason the wheat states adopted more technology than in the rice regions? Assuming constant prices for technological inputs in the production of wheat and rice across the ten agricultural states, the product market is observed as a result of its role in stimulating increased marketable surplus. In order to assess the role of prices in crop production, price movements of rice and wheat were observed over twelve years (1960 to 1972). According to the data in Table 2.7, average wholesale prices of the two crops tended to move together during this period, indicating a price increase between 1962 and 1967 that coincided with periods of drought and food shortage, preceded and followed by relative price stability. In fact, the average price of rice consistently exceeded that of wheat. It should be noted that government support prices, which are common in India as in many developing countries, do not alter the relative picture since they are included in these averages.[24] Hence, on the state level, one cannot impute the adoption of innovative technology in wheat producing states to an incentive originating in the product market.

TABLE 2.7

Average wholesale prices of wheat and rice (in Rs. per qu), 1960-1972

	1960	1962	1965	1967	1970	1972
Wheat	43.5	41.7	70.5	99.0	95.9	80.6
Rice	61.4	60.4	71.0	98.7	103.7	107.1

Source: various *State Statistical Abstracts*.

(4) Land Ownership

There is controversy in the extensive literature on tenancy and land reform as to whether ownership of assets gives farmers incentives to invest in land improvements and adopt technologies that require purchased inputs. A common generalization is that where conditions of tenancy or sharecropping prevail, increases in production disproportionately benefit the landowner, whereas the cost of production is disproportionately borne by the cultivator. Under these conditions, there is little incentive to incur debt in order to adopt new techniques whose outcome is often unknown. This was a favorite economic argument in favor of land reform, underlying many reforms that occurred in developing countries during the 1950s and 1960s. However, the evidence does not warrant generalizations of this nature since regions differ in their particular farming systems and in the characteristics of their rural institutions. In addition, it has been pointed out that the nature of the tenurial agreement is relevant in the decision by the cultivator to innovate, as is the distinction between pure tenants and tenant-owners (Newbery 1975, Scandizzo 1979, among others).[25]

In the Indian context, the evidence pertaining to the role of tenure in the adoption of technology is extensive and mixed. Parthasarathy and Prasad (1978) found that owners of land are more likely to adopt HYV seeds than tenants as a result of the risk factor. Bahduri (1973) similarly found lower rates of adoption among tenants but attributed this to the prohibition by landlords of innovative practices on tenants land. Vyas (1975) on the other hand, cites studies showing that the adoption rate in India is the same among owners and tenants with respect to HYV of wheat, and that in some regions tenants used more fertilizer per hectare than did owners.

The relationship between land ownership and innovation in the rice and wheat states of India is investigated by a state level comparison of the percent of area worked by owner cultivators. It is found that the wheat producing states of Punjab and Haryana had the lowest percent of owned farms of marginal and small size, and among the lowest in the medium and large size categories. Many of the low growth rice producing states had higher proportions of owner cultivators, which would imply greater incentive for technological adoption. Yet, that is clearly not the case.

TABLE 2.8

Area owned and operated as a percentage of total operational holdings, by category of holdings, 1971

(in acres)	Marginal 1.5-2.5	Small 2.5-7.5	Medium 7.5-25	Large over 25
A.P.	89.3	87.0	83.7	80.7
Bihar	99.6	99.7	99.7	99.6
Haryana	79.6	81.2	83.7	86.8
Karnataka	89.6	90.0	88.7	78.3
Kerala	85.7	81.1	77.7	82.1
M.P.	98.1	98.1	97.8	97.1
Orissa	89.3	90.7	92.8	93.0
Punjab	79.7	80.0	82.9	85.8
T.N.	91.7	90.2	91.3	92.0
U.P.	95.9	96.5	96.9	95.9

Source: various tables in I.J. Naidu (1975), *All-India Report on Agricultural Statistics*, New Delhi: Department of Agriculture.

(5) Farm Size

The size of the operational holding may be a critical determinant of the rate and extent of adoption of innovative technology. A myriad of considerations are associated with the issue of farm size, such as access to credit, to scarce inputs, to information, as well as the capacity to bear risk, so that the relationship between farm size and innovation is difficult to observe in isolation. The principal impediment to innovation by smaller farms is the fixed cost of inputs. The literature on the subject is mixed: some suggest that both the rate and extent of adoption are slower in the case of small farms in specific cases and locations (in Africa: Weil 1970, in South Asia: Binswanger 1978).[26] On the other hand, in his summary of the large body of literature on the subject, Ruttan (1977) contends that farm size is not a constraint to the adoption of HYV of grain. He furthermore points out that while a difference may exist in the initial stages of technological introduction, the lag identified with the small farms disappears within a few years.[27]

Although it has been claimed that HYV seeds are scale neutral and that high yields can be realized on any size farm (Srinivasan 1971, Sen 1974, and Singh 1979, among others),[28] the supporting technology in the form of irrigation and machinery *do* exhibit economies of scale and consequently, those farmers whose land plots are of such a size as to reap the greatest benefits from the new technology might have more incentive to adopt it (this contention is supported in the case of tractors by Binswanger 1978, and irrigation facilities by Dobbs and Foster 1972, as well as Gafsi and Roe 1979).[29]

It is therefore hypothesized that in the wheat producing states, the nature of some supporting inputs were appropriate for the predominant farm size in the region. On the basis of several productivity indicators, the appropriate size of operational holdings undergoing innovation associated with the green revolution ranges from 7.5 to 25 acres, which according to the Indian government classification is the medium size farm category (Khurso 1973). According to the 1971 Agricultural Census, 48.5 percent of the cultivated area in Punjab and Haryana falls in this category, compared to 38.2 percent in U.P (Naidu 1975). There exist two technical inputs associated with the green revolution whose form and size was most appropriate for middle size farms: irrigation facilities and tractors. With respect to irrigation, there exists either surface (i.e., canal) or groundwater irrigation (i.e., tanks, tubewells and open wells). The former tends to be large scale, and according to Singh (1979), its benefits mostly accrue to the large farmers. Canals entail large scale state level public investment and are a significant factor in interstate growth differentials. Groundwater irrigation systems also exhibit scale economies since they require a large sunk cost which is uneconomical on small or fragmented holdings. A relatively new development in irrigation is the private tubewell which best irrigates farms between ten and twenty-five acres in size. Smaller as well as larger farms find it unprofitable to adopt tubewells. However, in wheat producing Punjab, Haryana and U.P., 44.1, 32.3 and 34.9 percent of irrigated area is covered by tubewells (Table 2.9). These values are larger than for any other Indian state, suggesting that the nature of the input made it suitable for the particular farms prevailing in those three states.

TABLE 2.9

Irrigation by source (as a percent of total irrigation), 1970-71

	canals	tank	well	tubewell
A.P.	46.6	34.7	13.0	1.5
Bihar	40.4	4.7	9.5	12.7
Haryana	59.2	neg	6.3	32.3
Karnat.	42.4	49.8	17.9	neg
Kerala	45.1	16.3	15.3	neg
M.P.	48.0	8.8	37.3	0.8
Orissa	70.1	20.5	1.8	1.6
Punjab	48.3	neg	6.8	44.1
T.N.	34.1	32.6	30.5	1.9
U.P.	34.5	3.9	21.6	34.9

neg.: negligible
Source: various tables in I.J. Naidu (1975), *All-India Report on Agricultural Statistics*, Department of Agriculture, New Delhi.

Although mechanization *per se* is not strictly necessary for the realization of the potential associated with the green revolution, Binswanger (1978) provides evidence that it does increase farm efficiency. However, significant mechanization tends not to take place on small farms since the cultivated area is too small for indivisible machinery to be profitably applied. Table 2.10 contains data pertaining to an indicator of the mechanization process, namely the increase in tractor ownership. Clearly Punjab and Haryana exceeded the remaining states with respect to the degree of mechanization (298 and 279 percent, compared to an average increase of 146 percent in the rice producing states). Since this data fails to differentiate between types of tractors, it is relevant to introduce the research of Goetsch (1977) according to which the tractors widely used in Punjab and Haryana are appropriate for farms smaller than 25 acres, thereby contributing to widespread mechanization in those areas. They are tractors with small horsepower engines, which are furthermore locally produced and thereby adapted to local conditions (this aspect of mechanization is discussed at length in chapter five). The tendency in other Indian and Pakistani regions, as Goetsch points out, is to adopt large, often imported machinery that is economic only on large farms. Goetsch's findings point out the tendency of farmers in Punjab and Haryana to adopt mechanical inputs appropriate for the prevailing farm size.

TABLE 2.10

Tractor ownership

	1961	1966	1972	%change 1966-72	%change 1972-77
A.P.	1762	2911	6300	116.4	76.2
Bihar	1520	2132	5600	162.6	87.5
Haryana	na	4850	18400	279.4	65.2
Karnat.	981	2295	5700	148.4	45.6
Kerala	276	418	1500	258.9	13.3
M.P.	2025	2513	5000	99.0	202.0
Orissa	194	667	1800	169.9	-22.2
Punjab	7866	10646	42400	298.3	57.3
T.N.	1387	3278	5400	64.7	20.4
U.P.	7139	10139	27600	172.2	158.3

Source: various *State Statistical Abstracts*.

In two of the wheat producing states, Punjab and Haryana, the application of "technology appropriate to size" resulted in high productivity (as measured in yields per hectare) not only for their major crop, but also for the minor crop, rice. Upon comparison of the productivity in these states with U.P., it is clear that the latter has consistently attained a lower output, a phenomenon perhaps attributable to the difference in farm sizes. The distribution of *wheat* production in the U.P. is concentrated in the small farms, followed by medium size farms. In fact, of the irrigated area devoted to wheat production in the U.P., 2,255,000 hectares are covered by marginal and small holdings, compared to 1,352,000 hectares by medium farms and 278,000 hectares by large farms (Naidu 1975). This predominance of small farms, coupled with the use of the same seed and the prevalence of similar type of irrigation as in the Punjab and Haryana, resulted in lower productivity of wheat and rice. In fact, the average output of wheat in 1977-78 in U.P. was 14.3 quintals per hectare, compared to an average 23.2 in Haryana and Punjab. A similar difference is noted with respect to rice: 10.6 and 29.8 quintals per hectare, respectively, for the two sets of states. Thus, although the major crop is the same, the U.P. wheat farms have not been as productive as their counterparts in Punjab and Haryana, possibly as a result of the predominance of small operational holdings for which the prevailing technology is not appropriate.

Conclusion

The farmers in wheat states adopted more technology during the 1960s and 1970s than farmers in rice states, resulting in higher productivity of both wheat and rice in the wheat regions. This occurred as a result of the nature of the green revolution package of technologies which was appropriate for the prevailing farm size in Punjab and Haryana. The relationship between farm size and technology is also useful in the explanation of differences in adoption rates *among* the wheat producing states, Punjab and Haryana on the one hand and U.P. on the other. These findings do not imply that small farms are inappropriate for the cultivation of HYV rice. Indeed, there is ample evidence from India and across Asia of the lack of large economies of scale in rice farming systems and the prevalence and high productivity of family farms. Instead, it is simply argued that the green revolution package of technologies was appropriate for medium size farms during the 1960s and 1970s, and that this might be reproduced with a lag and with modification of inputs in other states at different times. Size of operational holdings is not *per se* a constraint to economic growth but may become so when viewed in conjunction with existing technical inputs. Economic growth can thus be stimulated either by adjusting the inputs to the prevailing farm size, as occurred in Punjab (and discussed in chapter five), or by adjusting the farm size, through various forms of land reform.

Notes

1. E. Denison (1974), *Accounting for U.S. Economic Growth 1929-69*, Washington: Brookings Institution, and John Kendrick (1961), *Productivity Trends in the U.S.*, Princeton: Princeton University Press.

2. The primary sector consists of agricultural and allied activities, as well as mining and quarrying (which is usually part of the secondary sector, although in this study, the classification used by the Indian government in the 1961 census was adopted). The secondary sector consists of manufacturing, processing, servicing and repairs and construction, whereas the tertiary sector comprises trade and commerce, transport, storage, communications, domestic and other services.

3. The Indian government classified economic activity into nine industries, the composition of which changed slightly in the course of 1961-1971, and was modified dramatically in the 1981 census to consist only of four categories: cultivators, agricultural laborers, household

industry and other. This classification applied to income as well as the labor force.

4. The Department of Statistics of the Indian government set guidelines for state statistical bureaus to follow in the compilation of state income. Specifically, it was suggested that the value added approach be used in the commodity producing sectors, the income approach in the non-commodity sectors, and when production extends beyond the borders of one state, a complex set of indicators to allocate totals to each state was suggested.

5. This brief introduction to the issue of technology will neither entail an analysis of biases and neutrality, nor the process by which innovation occurs in an economy (for a review of these issues, see Paul Stoneman (1983), *Economic Analysis of Technological Change,* London: Oxford University Press). The discussion of technology throughout this study is limited to its relative importance in growth as well as its social and economic impact.

6. Lance Davis et al. (1964), *American Economic History: The Development of a National Economy,* Illinois: R.D. Irwin, revised edition.

7. Denison (1974), op. cit., Kendrick, (1961) op. cit., R. Solow (1957), "Technological Change and the Aggregate Production Function" in *Review of Economics and Statistics* 39.

8. E. Denison (1967), *Why Growth Rates Differ: Postwar Experience in Nine Western Countries,* Washington: Brookings Institution, and U.N. Economic Commission for Europe (1964), *Economic Survey of Europe in 1961* (Part 2), Geneva: U.N. Publications.

9. For a compilation of writings on all aspects of appropriate technology, see Marilyn Carr (1985), *The AT Reader: Theory and Practice in Appropriate Technology,* London: IT Publications.

10. It is recognized that this approach is problematic insofar as it fails to distinguish between the first introduction of a green revolution input, such as HYV seeds, and the subsequent additions of the input to the production process. An estimation of the distinction between the primary and secondary application of technology has not been attempted in this study.

11. The compounded growth rates (Table 2.2) are substituted by the percent change in income so as to coincide with the percent changes in production inputs.

12. There is evidence that the western wheat regions of U.P. grew at a faster rate than the eastern rice regions (Sen 1974).

13. The combined income of these three states is obviously biased downwards because of the inclusion of U.P. The omission of this state would result in a greater income gap between the rice and wheat producing states.

14. Robert Evenson (1974), "Research Extension and Schooling in Agricultural Development" in P. Foster and J.R. Sheffield (eds.), *Yearbook of Education*, London: Evans Brothers, and John Villaume (1977), *Literacy and the Adoption of Agricultural Innovations*, PhD. dissertation, Harvard University.

15. John Gerhart (1975), *The Diffusion of Hybrid Maize in West Kenya*, Mexico City: Centro Internacional de Mejoramiento de Maiz y Trigo, and D.T. Jamison and Lawrence Lau (eds.), (1982), *Farmer Education and Farm Efficiency*, Baltimore: Johns Hopkins University Press.

16. Finis Welch (1978), "The Role of Investment in Human Capital in Agriculture" in T.W. Schultz (ed.), *Distortion of Agricultural Incentives*, Bloomington: Indiana University Press.

17. Rati Ram (1976), "Education as a Quasi-Factor of Production: The Case of India's Agriculture," PhD. dissertation, University of Chicago.

18. Wallace Huffman (1977), "Allocative Efficiency: The Role of Human Capital" in *Quarterly Journal of Economics* 91, and Todd Petzel (1976), "Economics and the Dynamics of Supply" PhD. dissertation, University of Chicago.

19. Mark Rosenzweig (1978), "Schooling, Allocative Ability and the Green Revolution," paper presented to the Eastern Economic Association meetings in Washington.

20. In fact, the literature on the relationship between credit and technological innovation cannot divorce itself from the question of farm size. See Bhalla's research (1979), which shows that in the Indian context, credit constraints were a greater concern for small farms than for large.

21. Lucio Reca (1983) "Price Policies in Developing Countries" in Johnson and Schuh (eds.), *The Role of Markets in the World Food Economy*, Boulder: Westview Press.

22. Robert Evenson (1977), "Cycles in Research Productivity in Sugarcane, Wheat and Rice" in Thomas Arndt et al. (eds.), *Resource Allocation and Productivity in National and International Agricultural Research*, Minneapolis: University of Minnesota Press.

23. On the basis of these findings, some believe that agricultural growth is stimulated by the raising of farm prices. This view was

challenged by Raj Krishna, whose argument rests on the disruptive macroeconomic effects of such a policy that would result from the very large increases in prices of agricultural output that would have to occur in order to bring about increases in output under conditions of relatively low supply elasticities, such as those that exist in most low income countries (R.Krishna (1982), "Some Aspects of Agricultural Growth, Price Policy and Equity" in *Food Research Institute Studies in Agricultural Economics, Trade and Development* 18/3).

24. See chapter five for further information on government procurement prices of foodgrains.

25. David Newbery (1975), "Tenurial Obstacles to Innovation" in *Journal of Development Studies* 11, and Pasquale Scandizzo (1979), "Implications of Sharecropping for Technology Design in Northeast Brazil" in A.Valdez et al. (eds.), *Economics and the Design of Small Farmer Technology*, Ames: Iowa State University Press.

26. Lumpiness of technology has been circumvented by the prevalence of hired implements in some parts of Third World.

27. See Jamison and Lau (1982), op.cit., as well as Parthasarathy and Prasad (1978), and Ruttan (1977) and for examples entailing a positive relationship between farm size and adoption.

28. See note 15, chapter one.

29. Thomas Dobbs and Phillips Foster (1972), "Incentives to Invest in New Agricultural Inputs in North India" in *Economic Development and Cultural Change* 21, and Salem Gafsi and Terry Roe (1979), "Adoption of Unlike High-yielding Wheat Varieties in Tunisia" in *Economic Development and Cultural Change* 28.

3

Agricultural Technology and the Labor Force

The rural population in developing countries has been increasing despite migrations to the urban areas. At the same time, industrial employment expansion has not been sufficient to absorb this increasing population. In addition, there is evidence of increasing landlessness among rural populations, particularly in South Asia and Latin America. All these factors point to the role of agriculture as chief employer and source of livelihood for Third World populations. Consequently, issues such as agricultural technology, mechanization, and rural industrialization are critical insofar as they determine the employment opportunities and labor displacement in rural regions.

During the 1950s and 1960s, development literature and policy was biased in favor of heavy industry, and unlike today, agricultural employment and rural productivity were not considered crucial components of the development process in agrarian countries. The two-sector models of the 1950s were based on a view of the relationship between agriculture and industry according to which agriculture provided the manpower, demand, and capital for the development of urban based industry. Development policies based on these models resulted in severe imbalances in growth rates between sectors and between regions. This in turn stimulated micro level studies which stressed the complexities of agriculture, such as rural factor markets, marketing, and farmer decision making. In addition, the focus turned to intersectoral relationships and growth linkages responsible for simultaneous growth in agriculture and industry (Mellor 1976, Johnston and Kilby 1975).[1] In addition, Third World governments and international agencies introduced employment oriented approaches to development with the dual purpose of increasing labor productivity and creating jobs (Seers 1970).[2]

This orientation towards rural employment stimulated studies of the relationship between agricultural innovation and the size of the labor force. This entailed the assessment of technologies on the basis of their relative employment potential, as well as the effect of employment oriented agricultural strategies on the growth of output and the consequent employment/output tradeoff.

This chapter contains an analysis of interstate variations in the effects of agricultural innovation on the total (male and female) labor force in India. Specifically, it focuses on changes in size and activity breakdown of the agricultural labor force that resulted from the application of green revolution technology. The discussion is limited to the two high growth states, Punjab and Haryana, because it is there that the green revolution technologies were most widespread and consequently the effects on employment most pronounced. Although the other eight agricultural states have experienced increasing agricultural innovation,[3] this innovation was sporadic in nature and scattered over several districts, and thus was of insufficient magnitude to warrant the inclusion of the entire state in the analysis.

In the course of economic growth, the labor force experiences a structural transformation: the proportion of labor employed in the primary sector decreases, coinciding with the decrease in the proportion of national income derived from agriculture (Kuznets 1966). The reasons why economic growth causes a decrease in *income* derived from agriculture have been discussed in chapter two, and are related to the decrease in the demand for agricultural *labor* in two ways: first, the increased consumer demand resulting from economic growth tends to be directed towards goods produced by the secondary and tertiary sectors, whereas the demand for agricultural products decreases because of their low income elasticity. Second, technological change, the main source of economic growth, tends to be labor displacing. Thus, the decrease in demand for agricultural products and the nature of technological change together result in the decreased demand for labor in the primary sector (Kuznets 1966: 113). Empirical evidence from scholars in the Kuznets tradition shows that all-India has indeed experienced this pattern of change (V.K.R.V. Rao 1983: 34).[4] Is this pattern recognizable on a state level, specifically in Punjab and Haryana where the agricultural sector experienced high rates of growth?

Green Revolution Technology: Biochemical and Mechanical

Literature

The effect of the green revolution on employment has been in dispute since the late 1960s (Bartsch 1977, Gibbons 1980, Griffin 1974, Yudelman et al. 1971, among others). Empirical studies have provided evidence in support of both the employment creating as well as labor displacing ramifications of cultivation with HYV seeds. There is little doubt that the intensive patterns of farming associated with the new seeds *did* entail an increase in man-hours of labor. Given that these new seeds mature faster, multiple cropping is enabled, and the additional weeding, land clearing, soil preparation and other field activities contribute towards increased labor use. In addition, multiple cropping entails repetition of non-field activities (such as product preparation, transportation, marketing) several times per year, resulting in further labor use. These agricultural practices constitute the "biochemical" aspects of the green revolution technology, on the basis of which Johnston and Cownie (1969) contend that the green revolution increases rural labor absorption. In the Indian context, this view is supported by Shaw (1971) and Billings and Singh (1969), who found that rural labor requirements in north India increased as a result of the new technology. Studies of Punjab (Chawla, Gill & Singh 1972, Johl 1973, Mellor 1969) yield similar results, pointing to the labor absorbing qualities of the green revolution. Empirical studies of other parts of the world support these findings: Barker and Cardova (1978) found that the demand for hired labor time unequivocally increased following the introduction of HYV in the Philippines, whereas in Indonesia the same result occurred due to a change in harvesting methods (Stoler 1977).[5] In regions where labor absorbing technology prevails, the *availability* of labor to satisfy the increased demands of multiple cropping is obviously crucial. It was found that the HYV can be too labor absorbing relative to the labor supply, so that labor shortages may in fact prevent the adoption of technology (Harriss 1972).[6] Hicks and Johnson (1974) found that abundant labor supply is crucial in the determination of adoption rates of some rice varieties in Taiwan.[7] The most common labor constraint is the peak-season scarcity associated with the peak activity periods of HYV technologies, which may result in hesitant adoption of HYV in regions characterized by low labor supplies and inadequate possibilities for seasonal migration.[8]

It is the mechanical inputs associated with the green revolution that tend to be labor displacing. Although mechanization is not strictly necessary for cultivation with HYV seeds, it has been found to increase yields under some conditions. The increased efficiency associated with the introduction of capital intensive, automatic agricultural implements such as threshers, seeders and tractors increases the output but simultaneously decreases the labor input insofar as planting, threshing and land preparation become mechanized. Binswanger (1978) reviewed evidence for South Asia and concluded that the use of tractors for ploughing and transportation does reduce labor time. Agarwal (1983) came to a similar conclusion, although differed in her findings pertaining to the *net* employment effect on workers.[9] However, the evidence with respect to combine harvesters, wheat threshers and maize shellers is unequivocal: they displace labor. This was found in Malaysia (Asian Development Bank 1978), Philippines (Illo 1983) and India (Billings and Singh 1970).[10]

Studies of north India also point to labor displacement resulting from mechanization. Bal (1974: 389) estimated that labor requirements on Punjabi farms were reduced by 16.5 percent with the first introduction of mechanical inputs. A World Bank study (1975, cited in Sadhu and Singh 1980: 123) indicated that marginal workers were affected by mechanization to a greater extent than main workers as their labor became dispensable following the substitution by capital inputs. Furthermore, an important work by Raj Krishna (1974: 211) pointed out that in Punjab, during 1969 to 1974, the labor displaced by mechanization far exceeded the labor absorbed from increased cultivation and irrigation. And finally, Dasgupta (1977: 251) found that the wider application of tractors and harvesters reduced the number of labor days needed per unit of land.

In the evaluation of the employment effects of mechanization, several considerations must be taken into account. First, the difference between the short and long run effects must be assessed. It has been argued that in the short run, mechanization displaces the labor used in ploughing and transportation, but that these effects are often offset in the long run because multiple cropping increases yields and therefore income which in turn generates employment. Second, the differing effects on small and large farms are studied. The evidence indicates that large farmers tend to mechanize more readily than smaller ones, not only because they tend to have more access to capital but also because the size of their farms warrants tractor use for efficient cultivation. Small farmers, on the other hand, tend to mechanize less and cultivate with higher labor use per acre. Third, the *net* effect of mechanization on employment is further complicated by the potentially offsetting intersectoral income and employment linkages which increase other forms of agricultural employment, such as the production of dairy and fruit products, consumer goods and agricultural inputs. This indirect employment effect in India was studied by Krishna (1975) and Mellor (1976), both of whom found that additional work time created outside agriculture by technical change exceeded the work time lost in agriculture. The above considerations complicate the picture with respect to the employment effects of mechanization and thus prevent generalizations on the subject.

There are then two aspects of the green revolution: biochemical and mechanical. The agricultural practices associated with these tend to have opposite effects on the labor force: labor absorption and labor displacement, respectively. The existence of these offsetting tendencies has made it difficult to assess the net impact of the green revolution on labor requirements, and the determination of changes in the size or composition of the rural labor force must certainly take them into account.

Agricultural Technology in Punjab and Haryana

What aspects of the green revolution technologies prevailed in Punjab and Haryana? The wheat HYV were introduced to both states at approximately the same time, yet the extent of adoption of seeds, as well as their supporting inputs, differed in the two regions. Table 3.1 shows the values of inputs associated with the green revolution (HYV seeds, irrigation, fertilizer consumption and tractors) in various years over the course of the 1960s and 1970s. It is evident from the data in this table that the diffusion of green revolution technologies progressed over time in both states, although Haryana consistently lagged behind Punjab. Despite differences in total cropped acreage, the percent of wheat land planted with HYV seeds is similar in the two states (compare, in percentages, Punjab's increase from 3.58 to 99.50 to Haryana's increase from 1.73 to 95.20 during 1967 to 1984).[11] However, this similarity fails to show up in the case of irrigation and fertilizer consumption. Irrigated area as a percent of total area sown was substantially greater in Punjab than in Haryana (58.8 and 37.8 in 1968-69, rising to 80.7 and 59.2 by 1980-81, respectively). Fertilizer consumption was greater in Punjab at all points in time, and expanded at a greater rate than in Haryana (during 1970-71 to 1980-81, Punjab's consumption increased from 40.3 to 133.2 kilograms per hectare, compared to Haryana where the change was from 17.3 to 42.0).

Even with respect to tractor use, Punjab's experience differs significantly from Haryana's (in 1977 Punjab had 66,700 units in operation, compared to Haryana where 30,400 have been enumerated). Although the percent increase over these years is similar, the level of tractors per HYV acre differs. In 1966, there were .18 tractors per HYV acre in Punjab, compared to .37 in Haryana. By 1972 and 1977 this difference disappeared and the two states have almost identical rates of tractor to HYV use (.025 in Punjab and .023 in Haryana, followed by .030 and .028 respectively). Although both states experienced a dramatic increase in tractor use, the increase in land under the new varieties of seeds has been even greater, resulting in a decrease in the tractor/HYV ratio. Therefore, the green revolution technology differed in the two states insofar as Punjab *initially* emphasized the biochemical aspects, while Haryana placed more emphasis on the mechanical. This distinction in the nature of the prevailing technology disappeared in the 1970s.

An explanation of the interstate variation in green revolution characteristics entails state level differences in demand and supply of mechanical inputs. One explanation for the difference in *demand* for inputs rests on the size of operational holdings. Towards the end of the 1960s, 30.7 percent of the area in Haryana was under operational holdings of less than one hectare in size, whereas in Punjab the corresponding percent was 42.4. Given that HYV seeds are scale neutral, they can be profitably adopted by the small farmers of Punjab. However, mechanical inputs tend not to be scale neutral and therefore are not cost effective on small farms. This interstate difference in farm sizes may explain why the green revolution in Haryana was relatively more mechanized during the 1960s than in Punjab. Changes in farm size occurred during the 1970s: according to the 1981 Agricultural Census, 32.2 percent of the area in Haryana was under operational holdings of less than one hectare in size, whereas in Punjab 19.2 percent belonged to this category, indicating a significant decrease from 42.2 percent observed a decade earlier (Statistical Abstract of Punjab 1984: 154-5). This trend may be attributed to the tendency among larger landowners to evict small tenants as the profitability associated with self-cultivation exceeded rent (this occurred in parts of many Asian countries, including Pakistan, Philippines and Indonesia; Agarwal 1985). The structural changes in Punjabi agriculture, consisting of an increase in area under holdings of two to ten hectares and a decrease in those below one hectare, lead to the increased mechanization witnessed during the 1970s. Similar structural changes did not occur in Haryana. The agrarian structure during the 1960s and 1970s is related to the pattern of demand for mechanical inputs and is partially responsible for the convergence in the nature of the green revolution in the two states.

TABLE 3.1

Various indicators of the green revolution

(i) HYV wheat seeds (in 1000 acres, & percent of total wheat land)

		Punjab	Haryana
1966-67	%	3.58	1.73
	abs	59	13
1971-72	%	73.06	67.79
	abs	1695	796
1975-76	%	91.38	88.30
	abs	2195	1087
1983-84	%	99.50	95.20
	abs	3107	1705

(ii) Irrigation (irrigated area as a percent of total area sown)

	Punjab	Haryana
1968-69	58.80	37.80
1973-74	72.36	48.68
1980-81	80.70	59.25

(iii) HYV land as a percent of total irrigated land

	Punjab	Haryana
1971	75.9	70.5

(iv) Fertilizer consumption (kg per cropped hectare)

	Punjab	Haryana
1970-71	40.3	17.3
1974-75	47.3	15.0
1983-84	133.2	42.0

(v) Tractor possession

	Punjab	Haryana
1966	10,646	4,850
1972	42,400	18,400
1977	66,700	30,400
percent change (1966-1972)	298.3	279.4
percent change (1972-1977)	57.3	65.2

(vi) Tractors per HYV acre

	Punjab	Haryana
1966	.18	.37
1972	.025	.023
1983-84	.030	.028

(Continued)

Table 3.1 (cont.)
Source: (i), (ii) and (iv): various tables in Fertilizer Association of India, Statistics
Division, *Fertilizer Statistics*, various tables from various yearly volumes
(iii): Naidu (1975), Table 9.31.
(v): Government of India, Central Statistical Organization, *Statistical Abstract of India*,
various tables from various yearly volumes.
(vi): calculated from above.

The *supply* of mechanized inputs is also crucial in the understanding
of Punjabi mechanization. Punjab is an important producer of agro-inputs.
During the 1960s, the nature of mechanical inputs supplied to the market
was altered. Pumpsets, automatic threshers and tractors became smaller in
scale and therefore more appropriate for local conditions, and thus enabled
smaller farmers to profitably mechanize (this aspect of Punjabi
mechanization is described in detail in chapter five).

Together, the size of the farms (which affects farmer demand) and
the change in the nature of mechanical inputs (which affects supply) may
explain the relative difference and subsequent convergence of the nature of
the green revolution in Punjab and Haryana.

How does the nature of technological innovation influence the size
and activities of the agricultural labor force? It is hypothesized that
Punjab's agriculture was more labor intensive during the 1960s than
Haryana's, and that this difference diminished by the end of the 1970s.
Although both states experienced intensive use of HYV of seeds as well as
mechanical technology so that the labor absorbing and displacing
tendencies both existed, *the net effect of these tendencies was not zero*. In
all likelihood, in Punjab net positive labor absorbing tendencies prevailed,
whereas Haryana's agriculture was characterized by net labor displacing
tendencies. An empirical analysis of the primary labor force will indicate
whether this in fact occurred.

Empirical Evidence: The Labor Force

Data Constraints

The measurement of labor force changes, across states and over
time, is based on data provided by the Indian census surveys of 1961, 1971
and 1981. As a result of the publication of the 1981 census, changes in the
labor force associated with a period of high agricultural growth as well as
the more recent low growth can be assessed.

Among the shortcomings of Indian census data, perhaps the most
frustrating are the changes in classification of the labor force that occur

during the census decades. These changes, which make time series studies difficult, affected the definition of worker as well as the industrial classification of the labor force.

The distinction between main and marginal workers is critical in the changing definition of worker over the census decades. The former are employed more days of the year than the latter. An individual was classified as worker in the 1961 census if he/she worked the major part of the preceding season, or for the last two weeks preceding the enumeration. Hence, the term worker included both main and marginal workers. It included workers who worked full time as well as those who did occasional work in the recent past. The 1971 census defined workers as those who were engaged in "main" work, implying greater consistency and duration of employment. This resulted in the elimination from the work force of many workers employed part of the year or partial days, ones who would have been included if the 1961 definition prevailed. The 1981 concept of worker distinguished between main workers (those who worked most of the past year), marginal workers (those who worked for some part of the last year), and non-workers (those who engaged in no remunerative activity). Clearly, the 1961 census is not strictly comparable to the 1971 census, but is comparable to the 1981 census. The 1971 data are also comparable to those of 1981.

With respect to the industrial classification of the labor force, the most important difference between the various census decades occurred in 1981 when the nine-tier classification system was dropped in favor of only four categories of workers: cultivators, agricultural laborers, household industries and other. This change implies that only primary sector labor can be compared by state, over time. Furthermore, the marginal workers of the 1981 census are not subdivided by industrial classification, but only their total size is given, hence, a time series study of the industrial classification can at best compare estimated marginal worker values.

With respect to female workers, Bardhan (1984: 30) noted another type of data distortion. He stated that 28 percent of female workers shift between domestic and "gainful" work, depending on agricultural tasks related to the seasons. Therefore, the size of the female labor force depends on the time of year in which the census enumeration takes place.

Some sources of Indian data attempt to rectify the census shortcomings discussed above. Most important among these are the National Sample Surveys (NSS) and the Rural Labor Enquiries (RLE). In its 27th (1972-73) and 32nd Rounds (1977-78), the NSS adopted a somewhat modified definition of the labor force and obtained results similar to those of the census: with respect to overall participation rates, the NSS overview was less gloomy and with respect to the degree of casual labor, it was less optimistic. Although chapters three and four contain references to the NSS results, the analysis is based mainly on census data. Given the shortcomings described above, it is necessary to observe general trends rather than to draw specific conclusions.

This study of agricultural labor was conducted on the state level in the two high growth regions, thus the district values within each state have been aggregated into a single state value. Although this aggregation masks individual differences on the district level, it does highlight the trends in the state as a whole. Furthermore, some data used in conjunction with labor statistics, such as sectoral income, fertilizer and tractor use, are only available at the state level.

Participation Rates

The data presented in Table 3.2 show the participation rates of the total population in Punjab and Haryana as measured by the percent of the population that is employed (a breakdown by sex is included in Table 4.1, chapter four). Although the population of north India has experienced constant increases over the course of the 1960s and 1970s, the activity rates for the total population of both Punjab and Haryana dipped in 1971 compared to their 1961 and 1981 values. In the course of these two decades, the main and marginal activity rates in Haryana decreased (from 37.9 to 33.6 percent), whereas Punjab experienced a slight increase (from 31.1 to 33.0 percent). An observation of main workers only indicates a decrease in Haryana from 1971 to 1981 (30.4 to 27.9 percent) as well as in Punjab (32.9 to 29.0 percent). Furthermore, the difference between main workers in 1971 and 1981 indicates that there has been an increase in marginal workers during this period. The research of Krishnamurthy (1984), based on NSS data and discussed in greater detail in the following chapter, shows that most of these marginal workers are, in fact, employed in the agricultural sector.

Sectoral Shifts

The pattern of sectoral shifts of the labor force described by Kuznets and identified in all-India by V.K.R.V. Rao occurred in Haryana but not in Punjab, *despite* Punjab's high rates of growth based upon technological change (see chapter two).[12] Instead, the percentage of Punjab's labor force employed in agriculture increased from 1961 to 1971 (from 56.9 to

TABLE 3.2

Participation rates in Haryana and Punjab

		Haryana			Punjab	
	1961	1971	1981	1961	1971	1981
total lab.	2879	2654	3588	3466	3913	4850
			(4314)			(5508)
particip.	37.93	30.44	27.92	31.13	32.88	29.03
			(33.57)			(33.04)

note: the above values are in thousands (for lab=labor force) and the participation rates
are in percentages. The figures in parentheses represent the values for the main plus
marginal workers
Source: Government of India, Office of Registrar General and Census Commissioner,
Indian Census, New Delhi, 1961, 1971, 1981.

63.7, according to Table 3.3). This relative increase was also absolute: at
a time when activity rates were decreasing throughout the country, the
absolute number of workers in the Punjabi agricultural labor force
increased. However, by 1981, primary sector workers decreased to 59.2
percent of the total, although the absolute number continued to increase.
Thus, the relative share of workers in agriculture declined slightly partially
because agricultural workers were expanding at a slower rate than the total
labor force. Despite the decrease in the relative share of agricultural
workers, it was greater in 1981 than it had been twenty years earlier.

The sectoral shifts of labor in Haryana differ from those experienced
by Punjab and more closely conform to the pattern observed in all-India.
The steady decrease in the proportion of workers in agriculture (71.4,
66.8, 61.4 percent during the three census years) is not associated with an
increase in the proportion of industrial workers, but rather may be related
to labor absorption by the expanding service sector.

These data indicate that while the proportion of state income derived
from agriculture increased in Punjab and Haryana, the proportion of labor
employed in this sector differed in the two states. During the 1960s,
Punjabi agriculture was a net absorber of population, whereas the
proportion of primary sector workers in Haryana decreased. However, by
1981, the relative proportions of agricultural workers were similar: 59.2
percent in Punjab and 61.4 percent in Haryana.

TABLE 3.3

Labor force by sector in Haryana and Punjab

1961	Sector I total*	Sector I %total	Sector II total*	Sector II %total	Sector III total*	Sector III %total
Haryana	2056	71.41	340	11.81	481	16.71
Punjab	1973	56.92	619	17.86	872	25.16
1971						
Haryana	1774	66.84	313	11.79	576	21.7
Punjab	2491	63.66	520	13.29	903	23.08
1981						
Haryana	2202	61.37	na	na	na	na
Punjab	2863	59.15	na	na	na	na

*= in thousands
na= not available due to changes in census classification of workers
Source: Government of India, Office of Registrar General and Census Commissioner, *Indian Census*, New Delhi, 1961, 1971, 1981.

The Primary Labor Force

Table 3.4 contains data pertaining to the labor force engaged in various agricultural activities during the two decades under study. The experience of the two states with respect to their agricultural labor force differs. The absolute number of cultivators in Punjab increased from 1961 to 1981. Given the change in definition of worker over this time, cultivators employed as main workers must have increased sufficiently to offset the elimination of marginal workers from the enumeration. This consistent increase in cultivators failed to occur in Haryana, where a relatively large decrease is observed during the 1960s. The 1981 data do not indicate a return to the 1961 number of cultivators.

Although both states experienced a dramatic increase in agricultural laborers, more were absorbed in Punjab than in Haryana. A large component of agricultural laborers are seasonal workers, for whom demand exists only during certain peak agricultural seasons. These are often migrants from other districts or states such as neighboring U.P. and Bihar, although some come from as far as Kerala to partake in the increase of work opportunities arising from the new technology.

TABLE 3.4

The composition of the agricultural labor force in Punjab and Haryana

| | Punjab | | | Haryana | | |
	1961	1971	1981*	1961	1971	1981*
cultiv.	1603	1665	1758	1839	1303	1613
%	81.0	67.0	61.4	89.4	73.5	73.3
agri. lab.	335	787	1105	199	430	588
%	17.0	32.0	38.6	9.7	24.2	26.7
allied	38	39	na	19	41	na
%	2.0	1.6	-	.9	2.3	-
total	1976	2491	2863	2057	1774	2201

cultiv.=cultivators, agri.lab.=agricultural laborers,
allied=workers employed in allied activities.
* refers only to main workers
na=not available due to changes in census classification of workers
Source: see Table 3.3

Conclusions

The state variations in the nature of the green revolution technology during the 1960s is reflected in the agricultural labor force and the changes it experienced. The relatively biochemical phase of Punjabi agriculture coincided with an increase in absorption of agricultural workers. Similarly, while Haryana's prevailing technology was relatively more mechanical in nature, agricultural labor displacement occurred. During the 1970s, when interstate variations in agricultural technology decreased, the proportions of agricultural labor also converged.

The differing effects of HYV on the structure of the labor force in Punjab and Haryana has implications for agrarian policy throughout all Indian states undergoing agricultural innovation, although the rice regions inevitably have a different labor/mechanization experience. Under conditions of large population density and pressure on land for the provision of livelihood, the encouragement of either biochemical or mechanical aspects of innovative technology is an important policy consideration. Although most determinants of the nature of innovations entail private considerations and decisions (such as access to capital, quality of land, size of operational holding), it is nonetheless possible to exert some public control on the rate and nature of technological change, using such limited tools as specific taxes or subsidies, as well as adjustment of exchange rates, encouragement of domestic production of farm machinery, etc. This intervention may be necessary as a result of the trade-off between the short term and long term costs and benefits of labor saving technological change. A short term consideration is the preservation of jobs and the maintenance or improvement of living standards in rural areas. However, in the long run, it is in the social interest that labor productivity be increased, and with it farm incomes. Usually this requires that capital be substituted for labor. Thus the output/employment consideration is essentially a short/long term issue and must be judged as such by policy makers. The decision to implement controls which will affect the rate and nature of mechanization must be part of the overall short and long run policy objectives. These must furthermore consider the non-farm activities that are created as by-products of agricultural growth, which may offset the loss of work time due to the increased use of capital. In the case of the diverse Indian states, government's role in the choice of technology is facilitated by the high degree of decentralization which enables specific conditions such as labor scarcity or the prevalence of large farms to be taken into account.

Notes

1. Bruce Johnston and Peter Kilby (1975), *Agriculture and Structural Transformation: Economic Strategies in Late-Developing Countries*, New York: Oxford University Press.

2. Dudley Seers (1970), *The Meaning of Development*, ADC Reprint, New York: Agricultural Development Council.

3. As discussed in chapter two, the new technologies were first introduced in the mid-1960s as part of an intensive package (IADP) on the district level in most states. The most successful districts where located in A.P., T.N., Punjab and Haryana. Since the mid-1970s, when rice varieties became successfully adapted to Indian conditions and rice production took off, the diffusion of innovative technology became more even across states.

4. Balasubramanyam (1984: 46) contends that structural transformation of the Indian economy was so insignificant that it cannot be argued that modern economic growth, as described by Kuznets, actually occurred. This contention seems highly improbable given the convincing

the evidence presented by Rao (1983). See chapter two for a discussion of the structural transformation of the Indian economy.

5. R Barker and V.G. Cardova (1978), "Labour Utilization in Rice Production" in *Economic Consequences of the New Rice Technology*, Los Banos: International Rice Research Institute, and A. Stoler (1977), "Class Structure and Female Autonomy in Rural Java," Department of Anthropology Mimeo, New York: Colombia University.

6. Barbara Harriss (1972), "Innovative Adoption in Indian Agriculture- the High-Yielding Variety Program" in *Modern Asia Studies* 6.

7. William Hicks and Roger Johnson (1974), "Population Growth and the Adoption of New Technology in Taiwanese Agriculture," Working Paper in Economics #1974-E6, Columbia: University of Missouri.

8. Seasonal peak labor shortages may be overcome if neighboring regions peak at different times thus allowing temporary labor migrations. In addition, this bottleneck is sometimes overcome by temporary mechanization during peak seasons.

9. Agarwal claimed that although the use of tractors does displace some workers, it does not necessarily have a *net* labor displacement effect because it is family or permanent labor that is affected (ploughing is rarely done by casually hired labor) and their labor use is merely transferred to other tasks.

10. Asian Development Bank (1978), *Rural Asia: Challenge and Opportunity*, New York: Praeger Pub., and J. Illo (1983), "Wives at Work: Patterns of Labour force Participation in Two Rice Farming Villages in the Philippines," paper presented to the conference on Women in Rice Farming Systems, International Rice Research Institute, Los Banos.

11. This represents an aggregate over districts. There is evidence, for example, that Ludhiana district by far surpassed other Punjabi districts with respect to adoption of HYV.

12. The agricultural share of the workforce in all-India decreased from 69.5 percent in 1961 to 66.7 percent in 1981.

4

Female Agricultural Labor in High Growth Regions

Introduction

The question of how economic growth affects women in developing countries has been attracting much recent attention. For example, an increasing proportion of scholarly publications focus on the role of women, the United Nations created a Decade of Women to call attention to women's issues throughout international agencies, and women's groups around the world have become more vociferous in their demands for change. These trends reflect not only the underlying need to address a neglected subject, but also a new vogue. According to Rogers (1980: 9), "women have been discovered!" It is still unclear whether this will translate into significant contributions to the improvement of the lives of women in developing countries, but the assessment of their present condition is quite clear. Throughout developing countries, there is evidence of varying degrees of gender-based inequalities, such as access to resources, skills and the means of production, which are dynamic and change in response to stimulus. One of these stimuli is the onset of rapid rates of economic growth. Following the pioneering work of Boserup (1970) on the effect of modernization on women's lives and work, numerous studies have identified trends present throughout many developing countries: as a result of the technology which underlies "modern" production, the female contribution to the market economy has decreased, while the contribution to subsistence production, labor intensive activities and domestic chores has increased. By analogy, male participation in mechanized or modern sector activities increased. In addition, overwhelming evidence seems to indicate that the increased burden of activities in women's lives was not accompanied by analogous changes in their remuneration. With respect to rural females, global trends indicate that as a result of changes in the quality and quantity of activities performed by females, there has been a decrease in their control of household income and, quite often, in their access to consumption. It seems therefore that economic growth, with its source in technological advance, is not always beneficial to women.[1]

At the same time, research has emphasized the positive role of women in many facets of the development process. Specifically, it is recognized that females contribute to the supply of labor available for development both in the modern and subsistence sectors, the latter being crucial for family existence during times when the modern sector fails to provide livelihood. In addition, female economic activity is inversely related to fertility rates: studies of both less developed and industrial societies show that female wage employment may increase the costs of children relative to their economic benefits.[2] Furthermore, women working for wages tend to have contact with the world outside the home, and are therefore more exposed to modernizing ideas which they may transmit to their children.

These are generalizations which apply in varying degrees in certain situations and locations. Women are not homogeneous. Not only are there distinctions between females living and working in less developed and industrial societies, or urban and rural locations, but the degree to which some generalizations apply depend on class, wealth, age, even time of year and number of children.

This chapter addresses the issue of economic growth and its effect on one aspect of female existence: their employment outside the home. Specifically, it is a study of the nature of the work performed and the changing participation in the workforce of the female population in Punjab and Haryana, where high rates of economic growth occurred primarily as a result of the successful application of green revolution technologies to wheat production. Therefore, this study involves the effect of economic growth, green revolution technology and north Indian culture on the economic activities of the female population during 1961 through 1981. Some of the questions which arise in this study are: did agricultural innovations affect the male and female composition of the labor force equally, or did they result in the increased participation of one sex? In which economic activities was it culturally acceptable for women to work and how were these activities affected by technological change and economic growth?

Background
(1) Literature on Indian Female Employment

As part of the global trend, the study of female labor in India intensified in recent years. As late as 1959, the Indian government noted that there was no comparative data available to study the trends in Indian female employment (Chawdhari and Sharma 1961: 643). This regrettable situation was rectified by the publication of the 1961 census, and every census since then has given rise to new studies on female economic participation. These studies are varied in nature and scope: some limit themselves to presentation of data and its statistical analysis (for instance, Ghosh and Mukhopadhyay 1984, Krishnamurty 1984), others take a historical and cultural view (such as Chandna 1967, Chawdhari and Sharma 1961, Gulati 1974), and others yet compare districts, states as well as countries (Gulati 1975a, 1975b, 1984, Nath 1970). The most important trend observed in these studies is the general decrease in female labor force participation over the period 1961-1981:

> Detailed analysis of available data on employment in India points to the fact that in the recent decades there has been a drastic reduction in the number of women workers as well as in their work participation rates.
> (Ghosh and Mukhopadhyay 1984: 1998)

Studies further identified a negative correlation between the economic growth that characterized this period and female employment. According to Billings and Singh (1970: A173): "...with the economic development of an [agricultural] area, participation in farm work by women declines," however, rigorous studies of the reasons underlying this decline did not ensue.

(2) A Model of Economic Growth, Technological Change and Labor

In chapter two, Denison, Kendrick and Kuznets were introduced as pioneers whose research identified the critical importance of technology in the achievement of sustained high rates of economic growth. Furthermore, the evidence presented in chapter three indicated that economic growth, emanating from technological innovation, tends to result in a structural transformation of the economy which, among other

things, entails a change in the labor demand by sector. It was also
specified that this phenomenon occurs because the demand for workers in
agriculture *decreases,* both because technological innovation tends to be
labor displacing and because the demand for agricultural products
decreases with increases in income per capita (Kuznets 1966: 117-127 and
chapter three).

Based on the studies mentioned above, the broad outlines of a model
of technological innovation, economic growth and labor changes is
presented in Figure 4.1.

Figure 4.1
Technical change, growth and agricultural labor

tech. change ====> increase Y/P====> decrease labor in I sector

where Y/P is income per capita, and I sector is agriculture

In most situations, technological innovation is responsible for a large
portion of sustained long term growth, which in turn results in the
structural transformation of the labor force. For the purposes of this
chapter, the critical transformation consists of the decrease in the
agricultural labor force.

Although it is clear that the chain of causation occurs from left to
right on Figure 4.1, it can also be argued that economic growth affects the
degree of technological innovation that occurs within a given society (i.e.,
the greater the income of a society, the more funds will be available for
research and development of new methods of production aimed at
increasing that income still more). Furthermore, the labor force affects the
rate of increase in income (as workers become more educated, skilled and
healthy, their productivity increases, thereby increasing total income per
capita). Lastly it should be noted that the changing size and structure of
the labor force in turn affects the rate and nature of technological
innovation in the economy as the relative price of labor and capital
changes, in part the result of changing demand and supply conditions.[3]

The application of this model to India warrants some description of
the principal technology in use during the period under study, an
observation of the resulting growth, and a statement of the labor force

changes. It should be recalled from chapter two that Punjab and Haryana had the highest rates of economic growth during the 1960s and 1970s, as well as the highest level of income per capita. Furthermore, it was found that this growth emanated from the green revolution package of technologies. Lastly, in chapter three it was recognized that these technologies differed in Punjab and Haryana: in the former, they were *relatively* more biochemical in nature whereas in the latter, they were *relatively* more mechanical, and that this difference in technology did in fact result in different labor demands. The biochemical technology increased activities associated with multiple cropping methods, as well as land preparation, harvesting, weeding, etc. In Haryana, where the mechanical aspect of the green revolution prevailed, labor displacement occurred as a result of the introduction of tractors, mechanical threshers and pumpsets, etc., which displaced such activities as ploughing, manual threshing, transporting, etc. We now turn to an empirical assessment of how well this model fits the data pertaining to north Indian females. Did technological change, which resulted in economic growth, decrease female labor participation? Were male and female workers affected by innovation and growth in the same manner in the two states?

Empirical Evidence

Quantitative Section
(1) Participation Rates
The data presented in Table 4.1 show the percent of the population that is employed (participation or activity rate), by sex, in Punjab and Haryana.[4] From chapter three we know that the participation rates for the total population of both Punjab and Haryana dipped in 1971 relative to their 1961 and 1981 values. In the course of these two decades, the (main and marginal) activity rates in Haryana decreased whereas Punjab experienced a slight increase.[5] A comparison of main workers only indicated a decrease in Haryana from 1971 to 1981 as well as in Punjab.

The participation of males in the labor force dipped in 1971 in Haryana but not in Punjab. In 1961, the main and marginal male workers in Punjab accounted for 52.8 percent of the male population whereas in 1971, only the main male workers amounted to 52.8 percent. Assuming the existence of some marginal workers despite their omission from the census data, these percentages indicate an increase in total (main and marginal) male workers as a percent of the male population. By 1981, the main and marginal percentage increased to 54.2 whereas the main remained at 52.0. In Haryana, the pattern of male participation more closely followed that of the total population (described in chapter three), although the changes were not as marked: main and marginal male workers decreased from 52.2 percent in 1961 to 51.1 in 1981, whereas the main male workers increased from 47.3 percent in 1971 to 48.2 in 1981.

TABLE 4.1

Participation rates, by sex, in Haryana and Punjab

| | | Haryana | | | Punjab | |
	1961	1971	1981	1961	1971	1981
male lab.	2120	2542	3298	3184	3839	4598
			(3499)			(4790)
particip.	52.18	47.28	48.17	52.86	52.83	52.01
			(51.11)			(54.19)
fem lab.	759	112	290	282	74	242
			(816)			(717)
particip.	21.51	2.40	4.82	5.50	1.18	3.09
			(13.59)			(9.16)

note: the above values are in thousands (for lab=labor force) and the participation rates are in percentages. The figures in parentheses represent the values for the main plus marginal workers.
Source: Government of India, Office of Registrar General and Census Commissioner, *Indian Census*, New Delhi, 1961, 1971, 1981.

These relatively insignificant changes in the male participation over the course of two decades should be contrasted with the more dramatic swings in female economic activity. In Haryana, the main and marginal female workforce decreased from 21.5 percent in 1961 to 13.6 in 1981, whereas the main workers increased from 2.4 percent in 1971 to 4.8 in 1981. In Punjab, on the other hand, there was an increase over the years 1961 to 1971 in the main and marginal female workers from 5.5 to 9.2 percent, whereas the main workers increased from 1.2 to 3.1 percent during the 1970s. The initial difference in female participation in the two states is noted: 21.5 percent in Haryana and 5.5 in Punjab (1961).

This section is incomplete without a mention of the National Sample Survey (NSS) data which covers the period 1972-73 to 1977-78 and indicates a less dismal picture, partially as a result of the extended concept of work. According to the NSS, the participation rate among rural males in Punjab increased from 54.4 percent to 55.9 during this period, whereas in Haryana it increased from 48.9 percent to 50.5. With respect to rural females, Punjab experienced an increase from 26.8 to 27.6 percent, whereas in Haryana the data indicate a participation change from 28.8 percent to 26.6 (Krishnamurthy 1984: 2122). Although pertaining only to the rural population, the NSS data support the evidence of trends in total participation rates observed in the census data.

(2) Agricultural Labor Force and Its Male/Female Composition

Table 4.2 clearly shows the importance of the agricultural sector in the north Indian economy. In Haryana, the percent of the labor force working in agriculture was 71.4 in 1961, dropped to 66.8 in 1971 and dropped further still to 61.4 in 1981. However, the absolute numbers in 1981 were higher than in 1961. The pattern in Punjab was different: there was an increase in the proportion of the labor force employed in agriculture over the 1960s, from 56.9 to 63.7 percent. Although this number fell to 59.2 in 1981, the absolute number of primary workers continued to rise, indicating that the labor absorbing properties of the biochemical green revolution in Punjab did actually absorb workers in the primary sector. In Haryana, the mechanical aspect of the new technologies in Haryana predominated, so that the net effect was a lower proportion of workers in agriculture.

TABLE 4.2

The agricultural labor force in Haryana and Punjab

Haryana	1961	1971	1981
total*	2056	1774	2202
%total	71.4	66.8	61.4

Punjab	1961	1971	1981
total*	1973	2491	2863
%total	56.9	63.7	59.2

*= in thousands
Source: Government of India, Office of Registrar General and Census Commissioner, *Indian Census*, New Delhi, 1961, 1971, 1981.

Were female workers absorbed by Punjabi agriculture, and were female workers displaced from Haryana's primary sector? Table 4.4 shows that in Haryana, a dramatic decrease in female representation in agriculture occurred during the 1960s: from 31.5 to 4.2 percent of the labor force. The trend was similar in Punjab: from 7.5 to 0.5 percent. By 1971, female participation in the agriculture of both states was very low relative to previous years (as well as relative to other states, such as A.P. where females account for approximately 50 percent of the primary labor force) and the slight increase witnessed in 1981 did not significantly alter that fact.

(3) Male/Female Breakdown of Agricultural Activities

Workers engaged in agricultural activities may be subdivided into the following categories: cultivators, agricultural laborers and other (including mining and quarrying, livestock, forestry, fishing, hunting, plantations, orchards and allied activities).[6] Table 4.3 contains data pertaining to the percent of female or male labor in each category as a percent of all workers in that category, as well as all workers of that sex.

TABLE 4.3

Composition of primary sector workers, by sex, in Punjab and Haryana

Punjab

	1961		1971		1981	
	female	male	female	male	female	male
Cultivators						
a	46.8	46.2	5.4	43.3	10.3	37.3
b	8.2	91.7	0.2	99.7	1.4	98.6
Agricultural Laborers						
a	5.7	10.0	10.8	20.3	26.9	22.6
b	4.8	95.2	1.0	99.0	5.9	94.2
Other						
a	1.1	1.1	1.4	9.6	na	na
b	7.9	94.7	2.6	97.4	5.9	94.2

Haryana

	1961		1971		1981	
	female	male	female	male	female	male
Cultivators						
a	79.6	58.3	37.5	49.6	49.3	44.6
b	32.9	67.2	3.2	96.8	8.9	91.8
Agricultural Laborers						
a	5.5	7.4	25.9	15.8	22.1	15.9
b	21.1	78.4	6.7	93.3	10.9	89.1
Other						
a	0.3	0.8	1.8	1.5	na	na
b	10.5	89.5	4.9	95.1	10.9	89.1

(Continued)

Table 4.1 (cont.)
note: a=distribution of female (or male) labor force by category of worker (e.g.. female cultivators as a percent of total female workers).
b=females (or males) as a percent of total labor force in a given sector (e.g.. female cultivators as a percent of total cultivators).
na=not available
Source: see Table 4.2.

A breakdown of agricultural workers in Punjab indicates that the percentage of cultivators of the female sex measured merely 8.2 in 1961. This number decreased to 0.2 in 1971 and 1.4 in 1981, indicating that an insignificant proportion of Punjabi cultivators are women. Yet, with respect to the total female labor force, 46.8 percent of female workers (in 1961) worked as cultivators, identical to the proportion of male labor force engaged in this economic activity. On the other hand, in Haryana in 1961, 32.9 percent of the cultivators were women. However, by 1971 female cultivators decreased to 3.2 percent, rising slightly to 8.9 in 1981. Cultivation engaged 79.6 percent of the female labor force in 1961 (and only 58.6 percent of the males), and fluctuated during the period under study: 37.5 in 1971 and 49.3 in 1981.

Agricultural wage labor constituted a less important activity for female workers in 1961 in both states, absorbing 5.7 and 5.5 percent of the female labor force respectively in Punjab and Haryana. However, the evidence indicates that wage laborers are on the rise (in 1971, 10.8 and 25.9 percent of the female labor force were enumerated as agricultural laborers in Punjab and Haryana, further increasing to 26.9 and 22.1 percent respectively in 1981).

Allied agricultural activities, under the category "other," were insignificant both with respect to the percent of the female labor force they employed, as well as in terms of the proportional size of female participation.

It is therefore clear that female involvement in the agricultural activities of both states is very small in comparison to the male participation, and that it has, on the whole, decreased over the period under study.

(4) Female Labor by Sector

Trends in the shifts of the female labor force by sector are presented in Table 4.4. With respect to the sectoral distribution of female workers, both states experienced a decrease in the proportion of female labor employed in agriculture over the period 1961-1971 (Haryana: from 85.4 to

66.1, Punjab: 52.5 to 17.6, in percentages). A portion of this drastic
decrease seems to have been absorbed by the service sector, the
importance of which increased in both states (Haryana: from 7.3 to 23.2
and Punjab: from 22.0 to 66.2, in percentages). In terms of the proportion
of the females that it absorbs, agriculture increased in importance in 1981
again, although not quite to its 1961 levels. Unfortunately the trends in
service sector employment in 1981 cannot be observed due to the change in
census classification explained in chapter three.

　　　With respect to female employment as a percent of labor force by
sector, agriculture contained the largest percentage of workers in Haryana
(31.5 in 1961). The decrease in primary workers is again evident in 1971,
when Haryana female participation drops dramatically to a mere 4.17
percent. In 1971, the Punjabi female representation in the primary sector
reached its low point of 0.5 percent from an already relatively low 7.5. In
both states, female participation in industries and services dropped over
the course of the 1960s, but again, those two sectors cannot be observed
in 1981.

TABLE 4.4

Female labor force by sector

	Agriculture		Industry		Services	
	Haryana	Punjab	Haryana	Punjab	Haryana	Punjab
Females as a percent of labor force by sector						
1961	31.51	7.50	15.88	11.31	11.44	7.11
1971	4.17	0.52	4.47	2.50	4.51	5.43
1981	9.40	3.14	na	na	na	na
Distribution of female labor force by sector						
1961	85.38	52.48	7.10	24.82	7.25	21.99
1971	66.07	17.57	12.5	17.57	23.21	66.23
1981	71.38	37.19	na	na	na	na

na=not available, given the change in census definitions
Source: see Table 4.2

(5) Female Marginal Labor

　　　What has been the trend in women's marginal work over the period
1961 to 1981? Given the difficulty in estimating the marginal labor force as a
result of the change in definition of workers in the 1971 census, NSS data
were utilized by Krishnamurty (1984) to study the share of marginal labor

in the rural work force. He found that on average, India has been experiencing an increase in the female marginal labor force (from 31.4 to 32.3 percent during 1972-73 to 1977-78). The pattern in Haryana conforms to that of all-India (8.4, 16.3 percent, respectively in the two years under study), whereas in Punjab, the opposite trend is observed: marginal female labor decreased from 9.6 to 7.6 percent. (Krishnamurty 1984: 2125, Table 12).

(6) Rural/Urban Female Labor

A review of the rural/urban distinction of female workers was conducted by Krishnamurty (1984).[7] He found that participation rates of females in rural Haryana changed from 2.3 to 4.9 percent during 1961 to 1971 (a comparison with rural males indicates a large difference in participation: from 47.5 to 48.6 percent, respectively), whereas the urban regions experienced an increase from 3 to 4 percent during the same period. In Punjab, the analogous change in rural female workers was from 0.7 to 1.7 percent, indicating lower participation than in the urban areas, where it grew from 2.7 to 3.7 percent. A comparison with the trend in all-India shows that the rural participation rates in Punjab and Haryana are relatively low (in all-India, rural female participation rates were, in percentages, 13.4 and 16.0, compared to the urban 6.7 and 7.3) (Krishnamurty 1984: 2122, Table 2B).

The fact that Punjab is the only state in the Union where urban female workers exceed their rural counterparts was noted by Gulati on the basis of the 1971 census (Gulati 1975a) and by Krishnamurthy (1984) using NSS data. It is remarkable that this occurred in Punjab given its lack of concentrated large urban centers: in fact, despite Haryana's proximity to New Delhi's industry and services, the proportion of its female urban work force does not exceed the rural. The relatively high urban Punjabi female participation coincides with the evidence from Table 4.4 indicating a relatively high proportion of women employed in services (in 1971).

Qualitative Section
(1) Technology and Agricultural Activities

What agricultural activities were the female workers engaged in, and how were these altered by the green revolution technology? Agricultural activities can be divided by gender on the basis of skill, strength, or simply "innate talent." There exists widespread disagreement as to whether any of these are really predetermined and whether skill, talent and even strength is not merely a convenient rationalization for one sex to perform the preferable chores. In her excellent study of women in developing countries, Rogers emphasized that what is considered male work in some cultures is considered strictly female work in others, hence division of labor by gender does not transcend regional borders (Rogers 1980: 14). As early as 1937, Murdock studied some two hundred societies in developing countries and found that activities were by no means uniformly distributed, but that in some, an activity was exclusively masculine whereas in others it was predominantly feminine.[8] Mead (1962) found that even the ultimate western sexual determinism relating to childbirth and infant care is shared in fantasy by males.[9] On the other hand, cultures such as those of Mbuti pygmies or the inhabitants of rural Bali show negligible difference among gender roles.[10]

In the context of north India, the particular division of labor by gender is well defined and ingrained in the mentality of the rural population. Agricultural chores may be divided into routine and seasonal activities, performed by either cultivators or agricultural wage laborers. Routine activities, such as manuring, maintenance of cattle, and dairy activities, are almost entirely preformed by women. It is in the seasonal activities that females are less represented. Preparing land for tillage, as well as carting, are exclusively male activities. Sowing and irrigation are predominantly male, although some females engage in some types of sowing and irrigation.[11] Weeding and hoeing are performed by both sexes, although females tend to do the hand weeding, whereas males tend to use implements. Harvesting is also done by both sexes, although harvesting of crops is mostly in the domain of males, whereas harvesting of fodder and picking bajra, jowar, etc. is done by females. Threshing is also a mixed activity.

Which female activities underwent modification following the introduction of the green revolution? Palmer (1978) provided evidence in support of the following effects of HYV technologies in various locations across the Third World: males increased their effort in land preparation and harvesting, whereas women increased transplanting and weeding work, the application of chemicals, harvesting and processing. She also cited evidence from Sri Lanka, south India, Java and Bangladesh indicating that some tasks described above are easier to mechanize, and once that occurs female tasks become male tasks.

In Punjab and Haryana, the most important labor displacing innovations of the new technologies were the introduction of pumpsets for irrigation, wheat threshers, tractors and wheat reapers. Billings and Singh (1970: A172) point out that, as a result of the new forms of irrigation, as well as the introduction of tractors, females are not displaced in either state. However, those women in Haryana who were engaged in wheat threshing and reaping probably experienced a decrease in the demand for their labor. In Punjab, on the other hand, female labor was not used in these activities, therefore displacement did not occur. Thus, the predominantly female activities, such as manuring, cattle maintenance and dairying, were not affected by the green revolution technologies, but those Haryana females engaged in mixed-sex activities might have been displaced by innovations. This complements the quantitative evidence from the census, according to which the decrease in cultivators in Haryana was dramatic over the years of the green revolution (Table 4.3).

Discussion

Empirical evidence shows that during the period of the introduction and proliferation of the green revolution technologies in north India, a marginalization of the female labor force occurred, along with a decrease in their participation in the primary sector. Yet, given the nature of the activities that were changed with the green revolution, only some females, mostly in Haryana, can be said to have been displaced by the technology. Obviously then, technological change alone is not sufficient to explain the transformation experienced by the female labor force in north India.

Technology cannot be observed in isolation from the context in which it is used. In particular, the technology and the resulting economic growth must be placed within the particular cultural context of India, and only then can technology and economic growth contribute to the explanation of women's changing economic participation. Although Reddy (1975: 904) claimed that "fancy socio-economic variables" should not be introduced into an analysis of the female agricultural labor force, it is inappropriate to negate the importance of cultural factors in the determination of both employment possibilities for and attitudes towards female workers, and, in broader terms, the economic role of women in society.

Cultural Context of North India

The cultural factors which are relevant for a study of female employment consist of those aspects of Indian religion and social norms which dictate human behavior. Punjab and Haryana are inhabited predominantly by Hindus and Sikhs, and to a lesser degree, Moslems. All of these religions are characterized by female subservience to men, both in the home and in society at large. This subject has been addressed at length by Sharma (1978, 1980) as well as DeSouza (1975) and Khanna and Varghese (1978). Sharma's extensive research on the characteristics of male/female relations in North India identifies the "purdah mentality" as a critical determinant of female economic activity. This "purdah mentality" extends far beyond the wearing of the veil and acts to restrict female conduct by forcing women to remain not only physically covered but also invisible in their activities (Sharma 1980: 221). Furthermore, their use of "public space" is limited since it is considered unacceptable for women to be seen outside of their own property or to engage in contact with non-kin. These characteristics are relevant for female work participation insofar as they limit the ease with which females offer their labor in the market. As a consequence of the "purdah mentality," females in north India are restricted in their mobility and hence in the location and nature of the work they can engage in.

In much of Punjab, and in parts of Haryana, most of the landowning, farming population belong to the Jat caste. This traditionally cultivating caste is characterized by the image of a proud, hard-working and thrifty farmer. These traits may be viewed as the Indian counterpart of Weber's "protestant work ethic" and may partially explain why agriculture in Punjab is relatively more successful than in other parts of India.[12] However, these entrepreneurial farmers show a disdain for females that work for wages, as well as for the husbands and fathers that allow them to engage in such activities. Because the social status of the family is *inversely* related to female employment, males prefer not to admit that "their" womenfolk are part of the labor force: in the words of Ghosh and Mukhopadhyay, "...rural people often hesitate to assign the epithet of 'worker' to their womenfolk." (1984: 2002). This is especially true of the *Sikh* Jat (farming caste) which predominates in the Punjab: "Sikh communities...do not view with favor the participation of women in farmwork" (Billings and Singh, 1970: A169). Chawdhari and Sharma (1961: 646) also contend that in a Jat community, female workers are kept aside from male counterparts since the men of the region prefer not to work in the proximity of females. The attitude of Jat farmers, reflecting their cultural bias against women in agriculture, was also extended to wage laborers who tended to be non-Jats (Omvedt 1981: A152). The *Hindu* Jat caste, which predominates in Haryana, is less strict in its bias against female workers,[13] as evident by the higher proportion of females in the agricultural labor force in Haryana (31.5 percent) relative to Punjab (7.5 percent in 1961; see Table 4.4).

The division of housework among the sexes is another culturally determined aspect of the lives of Indian women. Essentially, the domestic division of labor follows the pattern observed in many cultures, according to which females are solely responsible for housework (cleaning, cooking, washing, etc.) as well as childcare. These responsibilities rarely decrease when the woman becomes a wage earner, hence wage employment constitutes an additional burden to an already burdened existence. Rogers points out that women in north India tend to have a larger work burden than males (1980: 156), a contention supported by Agarwal (1985: 120) who claims that in India, as in most of the developing world, women spend on average six more hours per day than males on non-leisure activities.

Response of Female Workers to Green Revolution Incentives

It is suggested that the combination of technological innovation and economic growth *in the particular cultural setting* described above results in a greater male response to the green revolution incentives in Punjab and Haryana. It is argued that male capacity to participate in the primary sector labor force exceeds that of females, regardless of whether the workers belong to the wage laborer households or the cultivating households. The argument in support of this is elaborated below.

Incentives associated with the green revolution became evident when increases in income and profits resulted from farming activities.[14] The dramatic change in the yields, and consequently income, affected labor both in the landowning households (large and small cultivators) as well as in the wage labor force (agricultural laborers) insofar as it provided an incentive to enter the high growth agricultural sector. In the case of agricultural laborers, incentives manifested themselves in the form of increased agricultural opportunities and wages, which resulted from seasonal labor shortages associated with green revolution farming.[15] These relatively high wages acted to entice workers away from less lucrative employment. With respect to the landowning households, the potential economic benefits of increasing acreage under the HYV seeds or adopting supporting inputs was great. Indeed, landlords in north India found it profitable to evict tenants or resort to "reverse tenancy" in order to increase the land under their cultivation (see Omvedt 1981).

Although these incentives were equally apparent to males and females, the males of the region were more likely to respond to them as a result of cultural factors constraining females. Several factors restrain the female response to employment opportunities. First, the issue of physical mobility is crucial in the determination of supply of female labor, especially in the case of day agricultural laborers. Women, in their role of homemakers tending home and children, and restricted as they are by the "purdah mentality," are less mobile and can therefore respond less readily to economic opportunity located outside the home or district.[16] Often seasonal labor requires travel and extended separation from family: a practice acceptable for males but unacceptable for females. Male workers, in their traditional role of breadwinners and providers, have the freedom to be absent from the daily household duties such as childcare and food preparation since these are outside their realm of responsibility. Physical mobility then is a luxury not accorded to females in this particular culture, and as a result of it, female wage laborers are at a disadvantage relative to men in their competition in the labor market.

The second factor in the determination of female response to employment opportunities is the loss of social status associated with paid work outside the home, *given* the above description of attitudes towards female workers. The issue of status is especially relevant in the case of the high income, cultivating households. In these, the role of women in decision making pertaining to agricultural issues is small, and their role in actual cultivation is even smaller, since women in wealthier landowning households are more isolated and withdrawn from the external economic system than wage laborers because, according to Sharma, they could afford the luxury of a very strict purdha practice (Sharma 1980: 217). Given the particular cultural norms, the social status of a family is inversely related to the economic participation of its females, hence if it is not strictly necessary, the choice of the women, as well as their families, is that they should be removed from extra-familial interactions. Therefore, their capacity to participate in work, even if they choose to do so, is constrained by attitudes and norms of what is considered proper behavior for women of their income level and class.

The third issue determining female access to paid work is their level of literacy and education, both of which are lower for females than for the males of the region (in 1981 in Haryana, 22.2 percent of the females and 47.8 percent of the males were literate, and in Punjab, 34.1 and 46.6 percent respectively; see Table 6.4). Although literacy and education are not strictly necessary for agricultural tasks, they are an indicator of general openness to the outside world and the degree of modernization, hence receptiveness to new ideas and techniques.[17] The lack of modernity associated with minimal exposure outside of the home, coupled with the "purdah mentality," acts to intimidate potential female workers and thus restrain their economic activities and decrease their capacity to engage in economic exchange and interaction outside of the home. Indeed, females that need to sell their crops or purchase agricultural inputs in the open market often engage the services of a male relative and thereby remove themselves from that level of economic activity.[18] Even if the women are literate and courageous, their ability to interact with non-kin for economic purposes is largely dependent on the attitude of those with whom they interact. In this regard, Sharma (1980) has described women's fear of ridicule and abuse from the men they must communicate with when engaging in market transactions. This situation, so different from that of females in parts of Africa (described by Boscrup 1970), increases the dependency of female workers on their male kin and may lead to the withdrawal of females from work involving open market transactions.

Thus, the relative capacity of male and female workers *on the whole* to respond to the green revolution incentives is largely determined culturally. These factors, combined with technological innovation, were conducive to the shift of females out of and males into the primary labor force. Technological change associated with the green revolution affects the female labor force in two ways. On the one hand, it increases the income and wealth of some households, which then enables the withdrawal of their female labor from the labor market. Economic growth is associated with the voluntary withdrawal of some females from the agricultural labor market, explaining partially the decrease in female participation in the primary sector even in Punjab, where the total primary workers increased (a situation not unlike that in other cultures, in which women choose to terminate employment when the need for their wages decreases). Under conditions of a tight labor market, voluntary withdrawal might occur as women concede their work to male kin.

On the other hand, technological change also affected the female labor force directly. Thus, the voluntary withdrawal from the labor force (resulting from income increases) must be differentiated from the involuntary displacement that occurred to some women who desired employment (despite the *overall* increase in the income per capita), yet whose economic activities were displaced. This displacement may occur either directly (by mechanization) or indirectly (as a result of male shifts into other activities when their work becomes obsolete or as a result of their loss in the competition against males). This displacement describes the experience of Haryana, where mechanization displaced some activities performed by both females and males, thereby increasing the competition for the work that remained.

Research by Ghosh and Mukhopadhay (1984) supports this distinction between involuntary displacement. They claim that the principal cause of the decrease in female labor participation as well as the marginalization of the female labor force is the involuntary displacement due to a bias in technology which causes "a smaller expansion of the sector where women have an advantage." Second in importance on an all-India basis is the "employment effect" which relates to situations in which there is little employment opportunity and men are the first to "grab what there is." They add that this effect is virtually insignificant in Punjab, but not so in Haryana, where they estimate it accounts for up to 48 percent of the decrease in the female labor force (Ghosh and Mukhopadhay 1984: 2002)

With this introduction of voluntary and involuntary withdrawal of female workers into the discussion, and the suggestion that the former is stronger in Punjab whereas the latter in Haryana, the model of technology and agricultural labor, presented at the beginning of this chapter, might be amplified as shown in Fig. 4.2.

Figure 4.2
Voluntary and involuntary shifts of agricultural labor

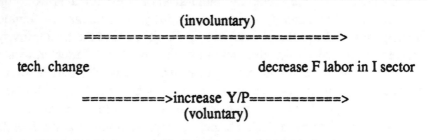

(involuntary)
=================================>

tech. change   decrease F labor in I sector

==========>increase Y/P===========>
(voluntary)

where Y/P is income per capita, F is female and I is agriculture

A discussion of the nature of withdrawals from the labor force cannot treat women as a homogeneous group. The distinction among women most relevant to the issue under discussion is the income category to which women belong, which in rural India is most clearly defined by land ownership. A drop in demand for female labor is experienced differently in landowning and landless households. In the former, female cultivators will tend to voluntarily withdraw from the labor force when the income of the family increases enabling a strict purdah practice. In the case of involuntary displacement among the landowning females, the increase in family income tends to offset the decrease in the women's income resulting in little change in *total* family consumption.

The experience of females belonging to landless households, who work as day laborers, is different. They engage in wage work out of necessity, their wages constitute a crucial part of the family income, and their displacement tends to result in a decrease in the family standard of living. Similar to the case of rural landless workers of the Middle East and north Africa, where purdah practice is irrelevant in determining the presence of low income women in economic activities, landless females of north India cannot afford to practice strict purdah nor to be withdrawn from society as their cultural and religious norms dictate. Instead, the increased marginalization of the female agricultural laborers that occurred during the 1960s and 1970s affected women and their households adversely.

The experience of the female workers in wheat growing states of north India might be contrasted with that of women in rice growing regions of the south, where HYV technologies also prevailed (albeit with a lag), and total income increased. Agarwal (1985) studied rural female labor by income category and found that participation of landless women in the casual labor force increased in both T.N. and A.P., whereas the experience of women from small cultivator households differed in these two rice states. Women from large cultivator households in both regions were unaffected by the green revolution since they never significantly participated in agricultural field activities.

Sectoral Shifts of Labor

The shift of women among sectors is mostly determined by the general demand and supply of labor. Western societies are known to have undergone sudden periods of increased demand for female workers in one sector, not as a result of specific policy aimed at improving the lives of females, but rather to fill the void caused by a decrease in supply. In Eastern Europe the period of reconstruction following World War II was characterized by the absorption of females (that were previously either unemployed or performing subsistence agricultural tasks) in the expanding manufacturing and services sectors, as a result of the low supply of men to satisfy this demand. However, as the structural transformation of the economy changed labor requirements and as the male labor force grew, females returned to agriculture which today is described as a predominantly female sector (Stipetic 1982).[19] Similarly, the manufacturing sector in the United States and western Europe experienced a decrease in labor during World War II as men withdrew from the labor force in order to serve in the army, and females were encouraged to fill the void. Following the return of men from the war, supply of labor in the manufacturing sector surpassed demand and to prevent a decrease in wages, governments opted for withdrawal of women from the labor force by creating obstacles to women's work: the closing of day care centers for children and "maternal deprivation" and "domestic science" theories proliferated, pressuring women to stay in the home (Rogers 1980: 24). In this case, the shift was not among sectors, but rather out of the labor force altogether. Both trends were observed in north India during the 1960s.

While the agricultural sector was experiencing high rates of growth, services and industries were not stagnating. In Punjab, household industries were on the rise. These are small scale industrial units, hiring few workers and often operating out of a home. Furthermore, services such as transport, storage, communication, construction and repairs existed in both urban and rural environments. In Haryana, the proximity of the metropolis of New Delhi provided the opportunity for service employment, including the domestic and cleaning service (which in India tends to be dominated by male workers). In addition, the economy of Haryana relies heavily on agro-industries for employment in the organized sector for processing, repairs, and trade and commerce.

The labor displacing nature of the green revolution that existed in Haryana during the 1960s forced males as well as females out of the agricultural labor force. These displaced workers, as well as new entrants into the labor force (resulting from population growth and migration), searched for work in non-agricultural activities. Displaced females competed with these males for employment opportunities in services, manufacturing and household industries. The net result, reflected by the census data, was that females left the labor force altogether while male participation in non-agricultural activities increased following their displacement from agriculture. It is suggested that females cannot compete with males for employment which males seek since cultural factors inhibit their capacity. Displaced males turned to service and manufacturing jobs, hence the women of Haryana, who participated in agriculture in the 1960s to a greater degree than in the rest of north India, were displaced.

This pattern of sectoral shifts did not occur in Punjab, where the agricultural sector was relatively labor absorbing due to the predominance of biochemical technology. However, the labor absorbed was essentially male. Given the profitability and high growth associated with the agricultural sector, males employed in other less lucrative sectors were attracted to the land. Their departure created a vacuum in non-agricultural sectors that could have been filled by displaced females, explaining the relative increase in female workers in services, industries and urban employment in general. High income in agriculture may therefore have contributed to the displacement of rural female labor in Punjab.

By 1981, the nature of the technologies (i.e., mechanical vs. biochemical) prevailing in the agriculture of Punjab and Haryana became similar (as discussed in chapter three), as did the decrease in the rate of economic growth (as discussed in chapter two). The increase in the proportion of females in agriculture that is observed in 1981 can be attributed to the same reasons, working in the opposite direction, as discussed above. The agricultural sector, no longer the fastest growing, ceased to provide the best incentives for males, increasing the opportunities for females to satisfy the labor requirements of agriculture. This indicates that changes in female labor coincide with changes in sectoral rates of growth. This inverse relationship between the income of a sector and the predominance of female workers within it is similar to a pattern observed outside of north India. Yousef (1974) found that in Mexico, Puerto Rico, Argentina and Brazil, female representation in manufacturing decreased dramatically within a few decades of mechanization and an increase in the rates of growth. In the United States, Patterson and Engelberg (1978) found the law profession to be characterized by high wages and little female representation whereas in the Soviet Union and other East European countries, the wages of medical doctors are relatively low in comparison to other professions, and the concentration of females within the profession is quite high.[20] Further support of this trend is the tendency of international agencies to prefer male participation in activities that promise to be lucrative. Rogers (1980: 142) discusses the failed efforts of the UN agencies to introduce cash crops by encouraging male involvement in regions where analogous activities were traditionally female.

Conclusions

The empirical evidence pertaining to female labor in north India indicates that female participation in the total labor force decreased over the course of 1961-1981 from an already low level relative to the Indian average. Female participation in the agricultural sector decreased in both Punjab and Haryana, although the technologies that prevailed in these two states differed with respect to their labor requirements. This decrease is explained by the interaction of economic growth and the technologies of the green revolution as it occurred in the particular cultural setting of north India, implying that the ramifications of technological innovation cannot be observed in isolation of the cultural context in which it is used. Technological change did not substantially alter the agricultural activities performed by females, although it did influence female labor insofar as it (1) changed the labor requirements in agriculture, which resulted in a change of the male/female ratio in all sectors, and (2) increased the income of the population, which altered female economic participation as dictated by cultural norms.

In addition to the importance of the cultural setting, the distinction among women workers by class is critical in the determination of female work. Females belonging to landowning households tended to withdraw from the agricultural labor force when the income derived from the green revolution increased, whereas among landless workers, the effect of the technological displacement predominated.

The class distinction is also relevant in a discussion of how changes in employment affect the quality of women's lives. A generalization pertaining to all rural women might claim that paid work increases the woman's personal independence, access to income and consumption, her self assurance as well as her status both within the family and society. This is obviously the case among some cultures for some particular types of work, such as the West African women engaged in marketing their agricultural produce. On the other hand, some scholars have generalized that the *decreased* female participation and increased domestication represents an improvement for society and for women as a whole: according to Billings and Singh (1971: A173), "... some females prefer not to work and a decrease in [their] participation can be viewed as an improvement in their status." A discussion of whether increased female participation in the workforce constitutes an improvement in the lives of women in north India must take into consideration at least the income group to which the female worker belongs, as well as the nature of the agricultural work she was involved in. The Indian women discussed by Billings and Singh must be those whose total family income increased during the green revolution despite their withdrawal from the labor force. However, the female day laborers who became marginalized or outright displaced and whose family income from other sources did not increase, did not enjoy an improvement in quality of life by any standards.

The relevance of these findings on technology, agricultural growth and female work extends beyond the context of north India to other developing countries, many of which are in the process of evaluating their development efforts and their position on female labor. If and when increased female participation becomes the goal of a government, then the ramifications of agricultural technology (such as the demand for labor and changes in income distribution) must be considered and therefore the choice of technology, if choice is possible, must be made appropriate for each particular cultural setting.

Notes

1. Changes in female employment and its ramifications have been addressed both in the western world as well as in the developing countries (Beneria 1982, Mickelwait and Riegelman 1976, Stromberg and Harkness 1978, Youssef 1974, among others). For research specifically on women and economic growth in developing countries, see Ahmed 1985, Lontfi 1983, Tinker and Bramsen 1976, Youssef 1974, among others.

2. The microeconomic theory of fertility states that the opportunity cost of childbearing is greater for females that work for wages than for the unemployed, therefore fertility rates tend to decrease with increased female employment (see, for example, Paul Schultz (1974), *Fertility Determinants: A Theory, Evidence and Application to Policy Evaluation*, Santa Monica: Rand Corp.). A recent example of the empirical evidence for this inverse relationship between labor participation and fertility is found in an ILO (1985) study entitled *Working Women in Socialist Countries: The Fertility Connection*, Geneva: UN Publications.

3. The model of a social structure which gives rise to technological change, which in turn promotes growth and lastly results in structural transformation of the labor force (which of course in turn affects the social structure) has been shown by Mandle, in his excellent study entitled *Patterns of Caribbean Development* (1983), to be at the crux of both

Marx's theories of capitalism, as well as Kuznets's theory of modern economic growth. Kuznets's structural transformation of the economy has been described in various chapters of this book, whereas the Marxian approach to growth and technological change is the focus of chapter eight.

4. The 1981 participation rates for all-India are, in percentages: 53.2 (males), 20.8 (females) and 37.6 (total).

5. Complete explanations of census terminology, such as the distinction of main and marginal workers, are found in chapter three.

6. Since there was a very slight difference between the 1961 and 1971 census use of the nine-tier system of classification, the labor force calculations were made on the basis of the 1961 system.

7. The category of rural workers consists of those employed in the rural areas, irrespective of whether their residence is rural or urban.

8. This is discussed in G.P. Murdock (1937), "Comparative Data on the Division of Labour by Sex" in *Social Forces*, 15/4.

9. See Margaret Meade (1962), *Male and Female: A Study of the Sexes in a Changing World*, Harmondsworth: Penguin Books.

10. See research by anthropologists Turnbull and Geertz as discussed in Rogers 1980: 16.

11. For example, females are exclusively responsible for sowing on ridges and planting of sugarcane, and with respect to irrigation, they partake in water control, and water changing among sub-plots. They do not make channels or maintain them (Chawdhari and Sharma 1961: 645).

12. See, for example, Goetsch (1977), as well as chapter five below.

13. Billings and Singh (1970: A169) divide the pre-reorganization (or Erstwhile) Punjab into districts, and note that those districts inhabited predominantly by the Sikh Jats are less likely to allow their womenfolk into the labor force than the regions which comprise present Haryana.

14. Numerable studies of the green revolution have shown that the profits derived from the new technology have been vast. See, for example, Dasgupta (1977), Day and Singh (1977), Frankel (1971), among others.

15. The seasonal change in agricultural wages, reflecting the changing demand for labor, has been identified by Day and Singh (1977). With respect to wage differentials between males and females, evidence indicates that it is by no means trivial, however, it will not be addressed in this chapter.

16. In Punjab, females as a percent of total migrants (during 1961 through 1971) is among the lowest in the country (1.08). Haryana females constitute 30.8 percent of the total migrants (see chapter seven and Premi

1982: Table A-6). This difference between the two states can be explained by the different attitudes towards female workers among the Hindu Jats and the Sikh Jats, as discussed above.

17. See the numerous articles in the 1974 World Yearbook of Education (P. Foster and J.R. Sheffield (eds), *Education and Rural Development,* London: Evans Bros Ltd.), as well as chapter two.

18. The degree of this problem of female dependency on males for market economic transactions has not been fully explored. An excellent recent study addressing this issue in a fishing village of Kerala, India was conducted by Gulati (1984).

19. See Vladimir Stipetic (1982), "The Development of the Peasant Economy in Socialist Yugoslavia" in Radmila Stojanovic (ed.), *The Functioning of the Yugoslav Economy,* New York: M.E. Sharpe.

20. See Michelle Patterson and Laurie Engelerg (1978), "Women in Male Dominated Professions" in A. Stromberg and S. Hawkins (eds.), *Women Working: Theories and Facts in Perspective,* New York: Mayfield Pub. Co.

5

Industrialization in a Highly Agricultural Region: the Case of Punjab

Introduction

India is ranked fifteenth among the nations of the world in terms of its industrial output. Production has increased at an average growth rate of approximately 5 percent per annum during 1970-1985. While most industries have contributed to this expansion, the growth has been particularly significant in chemicals, metal products, electronics and electrical machinery, transport equipment, petroleum products and power generation. Although the last three Five Year Plans have called for a spatial dispersal of industries and a diversification of products, neither the industrial progress nor the dispersal efforts are reflected in Punjab's development experience. In fact, extensive research on Punjab's current industrial capacity and its historical evolution has highlighted its unsatisfactory performance relative to other states of the Indian Union.[1]

Historical factors are the key to understanding the development of Punjabi industrialization. The most important of these was the British colonial strategy which incorporated India into world trade through the development of port areas and an extensive rail system. This combination of ports and railroads facilitated the movement of raw materials to Great Britain. A "division of labor" evolved within India, according to which industrial production was concentrated in the industrial belts surrounding major metropolitan areas while food production was relegated to Punjab.[2] This resulted in the development of certain regions, such as Calcutta and Bombay, as well as the retardation of industrial growth in regions such as Punjab.[3]

Despite the industrial stagnation that characterized British rule in Punjab, rates of economic growth steadily increased. From the data presented in chapter two, we know that this growth essentially derived from the agricultural sector, which traditionally accounted for over 50 percent of state income. In addition, the structural transformation of the economy, which was observed in all-India during 1951 to 1981, failed to occur in Punjab. The income derived from agriculture changed little during this period from its level of 55 percent, while the proportion of state income derived from industry increased from 10 to 17 percent. Since the late 1970s however, there has been a dramatic increase in industry, which grew by an annual average rate of 9.8 percent during 1979-1981 while agriculture expanded by a mere 1.5 percent (Government of Punjab 1983: 94). Therefore, although agriculture remains the predominant sector, the increasing importance of industries confirms that industrialization *is* occurring in Punjab.

This chapter describes the Punjabi industrialization process. It is a study of industrialization in a highly agricultural economy and an analysis of the role played by a growing agricultural sector in the development of some industries. Given the interaction of industry and agriculture, does the predominance of agriculture in the economy favor the creation of agro-related industries relative to others? In all-India, the agro-based industries account for a substantial portion of industrial output (60 percent in 1950, and decreasing steadily over time while the absolute value was increasing). Are they of similar importance in Punjab? Of specific interest for the overall development of the state is the role played by industries that produce inputs for agricultural production, such as the machine tools and agricultural implements industries. The quantitative and qualitative changes in these industries are described. Finally, the implications of Punjab's industrialization experience are addressed, especially with respect to the economic basis underlying the Punjabi separatist movement.

Mahalanobis and Mandle on Industrialization

The 1960s were characterized by a growing awareness of the interdependence between agricultural and industrial growth. These two sectors of the economy complement each other in the development process insofar as agriculture provides raw materials, foodgrains, labor, capital and demand for the industrial sector, whereas industry provides goods, employment and income to agriculture. A crucial intersectoral linkage consists of the flow of industrial products into the agricultural sector for production of agricultural products. These industries are usually classified as agro-input industries (industries producing inputs for agriculture), which together with agro-output industries (industries processing agricultural products) form the group of "agro-related industries."[4] These have proven to be an important factor in the industrial development of many countries that have a solid base in agriculture with a desire to increase exports of processed foods rather than raw materials and/or to decrease the import of those agricultural inputs that are necessary for its sustenance. The example of East European countries (Yugoslavia, Hungary, Czechoslovakia and Poland), with their large scale agro-industrial estates, shows how important this subsector of industry and agriculture can be for the national economy as well as for the balance of payments. Agro-output industries also contribute significantly to national income in several Central and South American countries.[5]

The development of agro-input industries, and its relationship to the growth process, has been studied by Mahalanobis and Mandle, whose research is especially relevant in the study of the particular industrialization process in the Punjab. According to both scholars, the crucial agro-input industry is the machine tools industry because of its contribution to the production of other goods and consequently to the growth of the economy.

Mahalanobis's research needs little introduction to students of Indian development. However, there are two aspects of his theories and policies that are often overlooked and are overshadowed by what he is best known for, namely the emphasis on heavy, large scale industry. The first is the emphasis on *industries which will enable production of other goods*, in other words, the machine industry.

> ...the key to industrialization lies in establishing
> the manufacture of...machine tools. Once this is
> done, everything can be gradually manufactured
> in India, mostly out of domestic sources.
> (Mahalanobis 1953: 82)

Thus, sustained and self-generating growth is made possible. Applied to agriculture, this implies that the local production of *industrial* goods for the local production of *agricultural* goods is of critical importance in achieving sustained high rates of economic growth.

A second aspect of Mahalanobis's theory that is relevant in the Punjabi context is his advocation of *production for internal needs* rather than for exports. His development model was based on a closed economy, and the Second Five Year Plan, designed according to this model, took little account of exports as a source of foreign currency and market for domestic production. It was suggested that Indian industrialization should primarily satisfy internal demand and stimulate diffusion and diversification of other industries. Although Malahanobis's emphasis on heavy industries has largely been discredited and proved inappropriate for the Indian context, the concept of production for internal needs is relevant in the ongoing debate on the relative merits of autonomous and dependent development.

In addition to Mahalanobis's theories, Mandle's research on the machine and tools industry is also relevant in the context of Punjab. According to this development scholar, in order for a country to overcome dependency on foreign imports and to become self-reliant, it is necessary to locally produce those goods without which an economy cannot sustain itself, namely machinery and tools. He writes

> The failure of (the machine and equipment industry) to emerge necessitates the importation of such capital goods as are needed from abroad...," thereby creating "...a condition of perpetual dependence on outside technology and capital.
>
> (Mandle 1984a: 113)

However, sufficient local demand must exist to stimulate the production of these goods, and the sector to which Mandle turns as a source of demand is agriculture ("...there is no doubt that a strategy of self-reliant growth will at least initially be dependent upon the agricultural sector..." (Mandle 1984a: 120)). This strategy will satisfy both the need for increased productivity of the agricultural sector (enabled by the introduction of more efficient machinery), as well as for local industrialization aimed at decreasing foreign dependence. However, two conditions must be met in order for this process to be effective: (1) the machines and tools must be appropriate for local conditions, and (2) in order for the demand to exist from the agricultural sector, government policy must ensure that the agrarian structure is such as to maximize farmers demand of machinery (Mandle 1984b).

Essentially, what is being discussed by Mandle and Mahalanobis is the relative merit of inward and outward looking development policies. The pros and cons of either type of development program have been amply discussed in the literature, and real world examples abound of countries opting for one or the other path and achieving successes and failures by both types of development efforts.[6] Mahalanobis and Mandle are essentially arguing in favor of creating *domestic markets for domestically produced goods*. A domestic policy of keeping up industrialization with the local demands for development achieves two goals simultaneously: it makes available inputs which otherwise would be lacking or would entail costly foreign currency, and secondly, it shields itself from the vagaries of demand by foreigners on which it has little control or influence.

Although the arguments above are usually made on a country level, some aspects of the discussion are relevant on a regional basis when there exists sufficient decentralization to enable development efforts independent of the center. This is the case in the Indian states. In an effort to study the economic growth of Punjab, the relative importance and development of the agro-input industries is observed and the relevance of Mandle's and Mahalanobis's theories is assessed.

Agro-Input Industries

Empirical Evidence of Growth

The levels of output and the growth rates of eighteen major Punjabi industries have been observed in order to identify the performance and relative importance of the agro-input industries. According to the classification of industries of the Indian government, agro-input industries consist of "agricultural implements and machine tools" and "fertilizer," which produce farm machinery and implements such as tractors, threshers, engines pumps as well as fertilizers: in other words, inputs for the production of agricultural output. "Steel re-rolling," "water-pipe fittings" and "nuts and bolts" are used in agricultural production (for example in the production and installation of various forms of irrigation) but will not be included in the agro-input category because a breakdown of output uses within the industry does not exist. Output levels, ranking, and average annual percent change of the eighteen industries are presented in Tables 5.1 and 5.2.[7]

An observation of growth during the 1970s indicates that the greatest expansion occurred in the production of rice shellers, sports goods, fertilizers, cycles and hand tools. Industries that experienced the least

TABLE 5.1

Punjabi industries by output (in Rs. x100,000), 1970-71 to 1980-81

industry	1970-1	rank	1975-6	rank	1980-1	rank
textile	2163	1	4536	5	18256	1
sugar	788	11	2069	11	2166	13
cycle	1725	6	5526	4	17898	2
sew mach.	399	13	892	14	1240	19
agri mach.	2161	2	4207	7	9808	7
hand tools	201	17	522	17	2043	14
steel re-roll	2146	4	7114	3	13371	4
pipe fitting	666	12	1140	13	1784	17
bolts,nuts	210	16	743	16	1862	16
hoisery	1655	7	3951	8	9307	9
wool textile	2153	3	4294	6	9406	8
cotton	2063	5	8412	1	10632	6
sports goods	235	15	856	15	3175	12
rosin, turp.	199	18	277	18	1886	15
fertilizer	1153	10	2684	10	12002	5
rice sheller	280	14	8063	2	15143	3
auto parts	1560	8	3326	9	4384	10
art silk	1330	9	1310	12	3755	11

note: textile= cotton textiles, cycle= cycle and cycle parts, sew mach= sewing machines
and parts, agri mach= agricultural implements and machine tools, steel re-roll= stell re-
rolling, pipe fitting= water pipe fittings, cotton= cotton ginning and pressing, rosin,
turp= rosin and turpentine.
Source: Government of Punjab, *Statistical Abstract of Punjab* 1984, Table 16.13

growth were water pipe fittings, sugar and auto parts. However,
observation of annual changes in output may be deceptive insofar as
industries that are by magnitude important will grow at slower rates than
smaller industries, thus the assessment of relative importance of industries
might include their ranking according to level of output. According to this
measure, by 1970-71, industries such as cotton textiles, agricultural
machinery, woollen textiles, steel re-rolling, and cotton ginning and
pressing had similar levels of output values and ranked highest among
Punjabi industries. By 1980-81, cotton textiles and steel re-rolling
remained of relative importance. In addition, new industries had come into
being and others underwent expansion so that cycle and cycle parts, rice
shellers, and fertilizers overtook the other industries at least with respect
to the value of their output. Two of these upcoming industries are
associated with agriculture, one as an input and the other as output. In

TABLE 5.2

Average annual percent change of Punjabi industries, 1970-71 to 1980-81

industry	%chg (1970-1980)	rank
textile	74.4	8
sugar	17.5	17
cycle	93.8	4
sew mach.	21.1	14
agri mach.	35.4	12
hand tools	91.6	5
steel re-roll	52.3	9
pipe fitting	16.8	18
bolts,nuts	78.7	7
hoisery	46.2	10
wool textile	33.7	13
cotton	41.5	11
sports goods	125.1	2
rosin, turp.	84.8	6
fertilizer	94.1	3
rice sheller	530.8	1
auto parts	18.1	16
art silk	18.2	15

Source: Derived from Table 5.1

fact, among the important industries, only cycle and cycle parts are in no way associated with agricultural production.

Industries processing agricultural output, such as cotton ginning and pressing, cotton textiles, and sugar, and rice shellers were of crucial importance to the Punjabi secondary sector, ranking first, fifth, eleventh and fourteenth in 1970-71 (in 1980-81, first, sixth, thirteenth and third respectively). Agro-input industries also expanded, although the relative size of agricultural implements and machinery decreased (dropping in rank from second to seventh place), while fertilizers increased dramatically (from tenth to fifth place). This decrease in production of agricultural machinery occurred despite the increased production of tractors all-India: a study by Agarwal (1983) indicates that tractor production increased 26 percent in 1980 (to one third the level of the USA!). Notwithstanding this *relative* decrease in the output of agricultural machinery, agro-inputs as a category of industries did expand in size over time as the Punjabi agricultural sector increased its use of inputs such as mechanical tools, implements, irrigation and fertilizer, etc. A closer look at the agro-input industries is thus warranted in order to fully appreciate the role of these industries in the Punjabi economy.

Qualitative Changes in Agro-Input Industries

The agro-input industries are crucial to the Punjabi economy insofar as they *enable* agricultural development, and are thereby indirectly responsible for the high rates of economic growth and levels of income in Punjab. Although empirical evidence points to a mediocre growth performance of the agricultural tools and implements industries during the first time period under study, as well as the machinery category of industries in the late 1970s, there is evidence that these industries underwent a qualitative change during the 1960s and 1970s. This transformation made input industries *appropriate* for local consumption, satisfied local demand for these inputs and thereby enabled high rates of growth in agriculture.

(1) Appropriate Agricultural Technology

Mechanized farming in Punjabi agriculture consists of the use of tractors, mechanical threshers, pumping sets, cutters, crushers, shellers and seed-fertilizer drills, the demand for which has been discussed in chapter two. Tractors are the most important among these, both with respect to the proportion of farmers budget allocated to them, as well as the percent of agricultural tasks they perform. Given that the average landholding in Punjab is approximately five acres in size, and 90 percent of the landholdings are under 25 acres,[8] the greatest demand is for medium size tractors: Randhawa (1974: 103) claims that a 15-30 horsepower engine is appropriate for these conditions since a smaller tractor cannot handle the various operations needed by the green revolution cropping patterns and intensities, whereas a larger engine has excess capacity on farms of this size.

Punjabi production of these relatively small engines was undertaken by some 400 farm machinery manufacturing units, characterized by small size and limited resources, as well as by several large producers. Punjab Tractor Ltd., one of the largest tractor producers in India, was set up in 1966 to produce primarily for local consumption. Another producer is Sawraj Ltd., concentrating in the category of 25-35 hp tractors. According to Gupta and Shangari (1980: 123), "..the growing demand for medium sized and lower value tractors due to land ceilings, [has resulted in] the Swaraj Tractors Factory adjusting its products". The Hindustani

Machine Tools Co. (HMT) also responded to the increased demand for tractors associated with the green revolution, and set up a tractor assembly plant as close to the source of principal demand as possible, in the already existing factory at Pinjore, Haryana (this region was part of Punjab prior to the 1966 reorganization of the states). The HMT produced the Zetor tractor, one of the most popular in use in Punjab. According to data collected by the Government of Punjab, the local and domestic tractor producers specialized in the production of small and medium size tractors (less than 30 hp) whereas the demand by large farmers for larger machinery was satisfied mostly by such foreign producers as Ford, Massey and Escort (Government of Punjab 1981: 51). However, this demand covered only 10 percent of the households.

Appropriate production for local conditions is also evident in the case of threshers. Ludhiana Thresher is the most important producer of threshers in the region, producing for local consumption as well as for other wheat regions. According to Goetsch, 95 percent of Punjabi wheat farms use these appropriate threshers, which are described as:

> ... small machines that were originally extensions of fodder choppers, a piece of equipment that has been a part of Punjab agriculture for decades. A housing was put around the chopper blades, then a fan introduced to produce the current machines that are capable of handling several quintals of grain per hour.
>
> (Goetsch 1977: 188)

In other words, they are produced with locally available technology and inputs, and suffice in capacity for the small farms prevailing in Punjab. High economic growth rates have been attributed to this appropriate technology: Goetsch (1977: 189) compared the Ludhiana threshers with those used under similar ecological conditions in Pakistani section of Punjab, concluding how in the latter the size of the machines is large, bulky, inappropriate and partially responsible for the lower growth rates of agriculture in Pakistani Punjab relative to its Indian counterpart.

With respect to water management, a trend has been observed in the proliferation of irrigation systems appropriate to local conditions. During British rule, large scale canal irrigation projects were introduced throughout the region. However, by 1980-81, 43 percent of the irrigated areas were under canal irrigation and 57 percent were under wells, tubewells and pumpsets (Government of India 1984: 78). Given the size of the farms, the resources of the small farmers and the water needs of the crops cultivated, the irrigation systems most appropriate for Punjabi conditions are wells, tubewells and pumpsets, which have proliferated in the last two decades. These irrigation systems have advantages insofar as they constitute a smaller capital investment by the farmer, the pumps are mobile and can be moved as the need arises, control of the flow of water is easier, and there are none of the problems associated with proximity and distance from the source as with canal irrigation. Goetsch describes the pumpsets used in Punjab as

> ... locally made and repaired, frequently being bought as kits by the farmer and assembled for him by a local mechanic who puts an engine together and adjusts it until it is running properly.
> (Goetsch 1977: 191)

Yet, this type of irrigation by no means proliferated in all wheat producing regions. A study of water management in north Indian wheat producing regions indicates that in U.P., for example, tubewells cover a mere 34.9 percent of the irrigated area.

Lastly, it is noted that seeds, although not strictly an industrial product, further demonstrate the prevalence of appropriate inputs in Punjabi agricultural production. Ninety percent of the area under rice and wheat cultivation in Punjab was cultivated with HYV seeds. These were made appropriate to local conditions such as soil, climate and taste *prior* to their widespread introduction to farmers in the mid-1960s. The Punjab Agricultural University, as well as other research stations, sponsored experimentation to ensure adaptation to local conditions of seeds originating elsewhere. In fact, this emphasis on "appropriate technology" in seeds has been cited by Mellor (1976: 55) as one of the reasons for the success of these crops in the Punjab amidst failure in other regions of India.

Thus, the agro-inputs industry in Punjab adapted itself to local conditions. This process occurred as a result of the increased demand from the agricultural sector, and specifically the demand for particular inputs which have proven profitable in the particular setting. The rapidity with which both industrial entrepreneurs (suppliers of agro-inputs) and capitalist farmers (consumers of agro-inputs) responded to the profitability associated with agricultural production was of critical importance. The supply of agro-inputs (seeds, as well as industrial inputs such as tractors, threshers, pumpsets, diesel engines) satisfied farmers' demand and stimulated further supply characterized by product modifications aimed at making the input even more profitable and appropriate. The principal ingredient of this symbiotic relationship was the response of both entrepreneurial farmers and industrial producers. What was the nature of this response and why did it occur in the Punjab?

(2) Entrepreneurial Response

When the green revolution offered the possibility of increasing outputs and consequently profits, farmers indebted themselves to acquire inputs necessary for the realization of these profits. On the basis of Johar and Khanna's (1983) comparison of input costs with output prices, there is little doubt that the profit motive determined the ready response of the entrepreneurial farmer.

For the purposes of this study, the response of the agro-industrial producer is of relevance. It emanated partially from the public and partially from the private sector. Most private local production of machinery occurred on a small scale in enterprises hiring less than five individuals. A common constraint was sufficient working capital, hence entrepreneurs supplemented their internally generated capital by collaboration with various public agencies that sought to, in the words of Franda (1979: 106), "encourage those who already conceive of themselves as entrepreneurs but who need funds and expertise." The government, through the DIC (District Industries Centers) and the IRDP (Integrated Rural Development Program) subsidized production of machinery by providing 15 percent of the capital investment, as well as subsidies on tools and interest free loans. In this way, the small scale private entrepreneur was able to produce the agricultural machinery.

In addition, the public sector was directly involved in machinery production. In 1966 the Punjab State Industrial Development Corporation was created to promote entrepreneurs in industries deemed insufficiently developed. One of these was the Punjab Tractors Ltd., that ranked third in size among 11 public enterprises and employed 982 workers. Johar and Chadna (1983: 237) studied state involvement in industry, and found that since 1974-75, this tractor producer experienced the largest profits of all Punjabi public enterprises, largely as a result of the demand by the agricultural sector.

Therefore, the entrepreneurs in both the private and public sectors in Punjab responded to the demand from agriculture. As Goetsch (1977: 191) points out, "under the pressure of a different demand structure, the local artisans of [Punjab] have mastered the ability to produce a key component of the mechanization process." Why did this response occur in Punjab? There exist at least two explanations. First, the Punjabi public and private sector had at their disposal information pertaining to agriculture that was lacking in the remaining states of India. Punjabi agricultural research facilities are among the most comprehensive and extensive in the country. Their agricultural know-how is based on direct observation and experimentation, and the close proximity of farmers enables rapid feedback between research and practice. The Punjab Agricultural University (PAU) conducts extensive research on all aspects of seeds, water management, fertilizer and pesticide inputs, and as part of its Machinery Research Program, explores alternatives for the development of machinery appropriate for the particular local conditions. Information from the PAU is readily available to farmers and entrepreneurs.

Another reason for the timely response among agricultural and industrial producers in Punjab is the existence of an entrepreneurial spirit which permeates shops, factories, and countryside. Within the Sikh community, emphasis is placed on innovation and adaptability, resulting in risk-taking of a degree not witnessed among other social groups in India. This is true of the agricultural caste, the Jats, as well as the urban and rural artisans, merchants and financial castes (the Kastias and Auroras). It is likely that these characteristics of the Sikh national character greatly influenced the production in the agricultural machinery and tools industry in the Punjab, specifically with respect to its innovative and adaptable nature.

Thus, information availability and entrepreneurial spirit resulted in prompt, extensive and effective response of the farmers and agro-input producers. The success of agriculture stimulated the further supply of inputs, and the success in the provision of agro-inputs contributed to the further success of agriculture. The green revolution technology provided the incentives for both farmers and agro-input producers, whereas the existence of agricultural information and entrepreneurial spirit enabled the agro-input industry to adapt. It seems therefore that intersectoral flows between agriculture and agro-input industries are strong, the ability to adapt these industry to changes in agricultural demand is evident, and the importance of agro-inputs for the Punjabi economy is indisputable.

Implications

Several implications arise from a pattern of industrialization characterized by the development and adaptation of the agro-input industries in response to agricultural stimulation. If observed from the point of view of Mahalanobis's and Mandle's research, a machine tools industry geared for local consumption is suggestive of an *enclosed economy* characterized by self-reliance and low levels of dependency on the outside economy. Mandle's research on the relationship between the machine tools industry and agriculture is especially relevant in Punjab, where farmers are demanders of appropriate machinery and tools, thereby raising their yields and stimulating the production of these industries. The increased local production in Punjab serves to decrease the dependency on foreign and out-of-state products which have often proven to be inappropriate for local conditions. Thus, the completion of the production/consumption cycle within one region reflects self-sufficiency. In addition, the economy with a solid base in the production of those inputs that enable its growth is is more likely to experience sustained economic development.

One consequence of a widespread perception of economic self-sufficiency is the movement for Punjabi political independence from the Indian Union. A review of the economic basis of Punjabi independence is particularly relevant in the mid-1980s in view of the recently intensified demands for an autonomous region, demands whose strength is evident by the political turmoil preceding and following the bloody invasion of the Holy Temple in Amritsar, such as the assassination of Indira Gandhi and moderate Akali Dal members, as well as the continuous state of terror among both Hindus and Sikhs in the region.[9] There are at least two economic considerations underlying these calls for separatism.

The first entails the apparent decrease in Punjabi dependency on both the Indian Union as well as the global economy emanating from the local and appropriate production of agro-inputs. The trend among Punjabi institutions and entrepreneurs to produce goods crucial for the economy resulted in a decreased need for domestic (Indian) or international imports of those inputs instrumental in achieving economic growth. Some Punjabis have viewed this as an indicator of self-sufficiency and of the ability to sustain its growth rates without reliance on the Indian Union.

The second factor that seemingly lends economic justification to an autonomous Punjab is related to the central government's policy with respect to interstate trade in agricultural commodities, specifically foodgrains. Given that Punjab's economy is predominantly agricultural and that foodgrains form the crux of this sector's production, foodgrain policy affects a wide spectrum of farmers and middlemen. The relevant aspects of Indian foodgrain policy are briefly presented below. In 1964, the Indian government introduced "agricultural commodity zones," across which trading was not permitted except under a public transfer program entitled "state trading." This enabled the central government to redistribute foodgrains from surplus to deficit states (as well as from rural to urban areas) using fixed prices ("procurement prices") aimed at providing a fair return to the suppliers while protecting consumers from unfairly high prices. In other words, prices were set so as to ensure the farmer some degree of profit, and the foodgrain consumer the ability to purchase. Given that Punjab is a surplus producer of foodgrains, this policy caused displeasure among some participants in the Punjabi economy. According to Narang,

> ...the farmers in Punjab felt that [procurement pricing policies] were not in their interest, nor good for India as a whole. They reasoned that if they could realize their full prices, they would have more to invest and could increase production still further.
>
> (Narang 1983:152)

This position had been adopted by some of the more vociferous supporters of an independent Punjab. In addition, the international dimension of this issue gives further support to this position. There is evidence that procurement prices paid the Punjabi farmers for wheat were lower than international open market prices, implying that a higher price for wheat could have been achieved if the Punjabi surplus were traded in the international market.[10]

The conditions described above have been interpreted as sufficient evidence that Punjab's economic prospects might improve if it did not bear the yoke of the Indian Union. Punjab does, after all, enjoy by far the highest level of income per capita, and it might be argued that even greater economic growth would be possible if it were allowed to maximize its exports. These conclusions reflect an incomplete view of economic relations within India and are debatable on several grounds.

First, the relative independence emanating from local production of agro-inputs for local markets cannot be observed in isolation of the spatial location of markets for the agricultural output that is so crucial to the Punjabi economy. The evidence from Punjab is that local agricultural production by far exceeds local demand such that out-of-state markets constitute an essential source of incentive for farmers. Indeed, Punjab's dependency on out-of-state consumer demand for its chief product, foodgrains, has been increasing since the mid-1960s. According to Punjabi government statistics, 85 percent of wheat production in 1970-71 was traded domestically or procured by the government, and in 1977-78 this proportion increased to 98 percent.[11] As a result of membership in the Indian union, Punjab has access to guaranteed markets as long as deficit production persists in the remaining areas. If and when food production increases to meet demand in deficit states, government regulations may relax sufficiently to allow Punjab's access to international markets where the foodgrain prices have been relatively higher.

A second argument against an autonomous Punjabi economy rests on the beneficial aspect of procurement prices to Punjabi farmers. In addition to providing a ceiling aimed at fair prices to urban consumers, procurement prices also act as a price floor when surplus might result in falling prices. Kahlon and Tyagi (1985) noted that when this situation occurred on several occasions during the 1960s and 1970s, wheat producers were protected from a potential decrease in prices.[12]

A third constraint to Punjab's economic self-sufficiency is the dependency on all-India markets for industrial inputs which in turn are used as factors of production in local industries, including agro-inputs. Punjab remains deficient in steel, iron and coal, some of which are essential for industrialization even at a minimum level. In fact, imports into Punjab of iron and steel from the Indian domestic market have more than doubled in the course of 1967 to 1976, and tapered off slightly by the early 1980s. Although these raw materials are partially used for the production of machinery, tools and agricultural implements *for export* both to other Indian states as well as out of the country,[13] they nonetheless indicate Punjab's dependency on the Indian economic union. A domestic market for these inputs has advantages over the international market insofar as it does not require foreign currency, and may be subject to government price intervention in a way that may suit Punjab.

A fourth point relevant to this discussion is the dynamic nature of Punjab's self-sufficiency in the production of agro-inputs. Although the *current* entrepreneurial and public sector response has been to mobilize resources and research in order to develop the most appropriate inputs, at the onset of the green revolution Punjabi dependency on global markets and research was extensive. It is not unlikely that the direction of international research and technological innovation influences international prices and production in such a way that Punjab's technological advances become outdated. In this case, Punjab might need to import technology and thereby experience temporary dependency which in turn points out the *dynamic* nature of Punjab's relation with India and the international community. This dependency may not be limited to the international research community, given the proliferation of research centers throughout India from which Punjabi producers may learn.

On balance, the economic arguments in support of an autonomous Punjabi economy are overshadowed by the benefits derived from access to markets. In this respect, Punjab's position and role within the Indian Union is similar to that of Slovenia state within Yugoslavia. Both regions are the wealthiest, enjoying the highest income and production capacity per capita. In both cases, the relationship with the remaining states is characterized by greater outflow than inflow of goods. However, the net gain derived from federation is positive in both regions and it is questionable whether their high rates of growth could be sustained without domestic markets. It is therefore concluded that the arguments for an independent Punjab cannot rationally be based on economic issues.

Notes

1. Studies on Punjabi industrialization are conducted by both the central government (originating from NCAER and the Economic and Statistical Organization) as well as the state government (originating from the Deptartment of Planning and Directorate of Industries). Non-governmental studies include: J.S. Khanna (1980, "Lagging Industrial Development in Punjab: An Analysis" in *PSE Economic Analyst* 2) and J.S. Sandhu and A. Singh (1983, "Industrial Development in Punjab" in P.S. Johar and J.S. Khanna (eds.), *Studies in Punjab Economy*, Amritsar: Guru Nanak Dev University), among others.

2. If measured by either industrial output or density of industrial workers, these belts are located in the following regions of India: Maharastra, Gujarat, M.P. and West Bengal.

3. This lagging industrialization has been addressed by J.S. Khanna in his analysis of the colonization process and the resulting regional

imbalances in the Indian economy (1983, "Lagging Industrial Development in Punjab" in P.S. Johar and J.S. Khanna (eds.), *Studies in Punjab Economy*, Amritsar: Guru Nanak Dev University, pp.89-94).

4. Although the literature on agro-industries is vast, and discussion arises as to exactly what is included in that group of industries, Bhattacharya's comprehensive classification seems appropriate in the Indian context. Broadly, agro-industries consist of the following: agricultural inputs (also called agro-inputs, consisting of mechanical, chemical and construction industries), consumer goods (demanded by agricultural population), agro-based industries (also referred to as agro-output industries which process agricultural produce), services, forestry based, mineral based, and handicraft industries. See S. N. Bhattacharya (1980), *Rural Industrialization in India*, New Delhi: B.R. Pub. Co., p. 193.

5. This has occurred as a result of the profit instinct of multinational corporations rather than domestic government efforts.

6. The discussion of import vs. export based growth is especially relevant in the mid-1980s given the recent experience of the NICs (newly industrialized countries) whose growth is essentially based on production for exports and therefore whose vulnerability resulting from recession and changes in the world demand is great. See articles in *South*, such as "Miracle Cure" (October 1985: 156) and "The Dragons Lose Their Fire" (November 1985: 121).

7. The Punjab Statistical Office collects output data for 22 industries, but four were omitted in this study because they were too insignificant in size, and/or no data was available on them until 1975-76.

8. See Naidu (1975) for data on Punjabi structure of landholdings. Punjab is one of the few Indian states in which land reforms were successful in both consolidating fragmented landholdings as well as decreasing the size of farms.

9. The radical segment of the Akali Dal party, as well as those too radical for its membership, have been engaging in terrorist acts in order to convey their message to both the moderate political leaders, as well as the general public. Harchand Singh Longowal, murdered on August 20, 1985, was a Punjabi leader who was assassinated in order to prevent a moderate agreement with the central government.

10. Two observations of the 1980s illustrate this trend: in 1982, the average annual wholesale price of wheat was US$ 154 per metric ton in Punjab, compared to 196 for Canadian (which moved quite closely with

the US and EEC prices), whereas in 1984 the numbers were 151 and 167 respectively (Zarkovic 1987a: Table 4)

11. Despite the drop to 77 percent in 1975-76, an insignificant quantity of crop was retained locally or exported (*Punjab Statistical Abstract* 1984: Tables 23.1 and 7.8).

12. However, it must be noted that the issue of procurement prices cannot be considered of critical importance neither with respect to its benefits nor its costs because the percentage of total production that is procured has been relatively low over the past two decades and rarely exceeding 10 percent of total foodgrains produced (Rao and Vivekananda 1979).

13. Exports to Indian states increased after 1971 (see Government of Punjab, *State Statistical Abstract*, Chandigarh: State Statistical Office, volume 1976: Tables 11.1 and 11.4, volume 1984: Table 23.2).

6

Economic Growth and Human Welfare

Introduction

Economic development is in great demand in developing countries. Although there is a concensus among most policy makers, development scholars and residents of developing countries that development is to be encouraged, disagreement arises as to exactly what development *does* entail and what it *should* entail. Those concerned with development issues have implicitly or explicitly defined the goals that a development program should strive for and how these objectives may best be achieved. Currently there is agreement among mainstream development scholars that the achievement of *economic growth*, a *decrease in poverty* and a climate of *political freedom* are reasonable development objectives for which to strive. However, this range of goals was not always prevalent. The 1950s and 1960s were characterized by a focus on growth maximization at the virtual neglect of other issues, reflecting a unidimensional approach to development that proved to be unsatisfactory. Experience in Venezuela and South Africa during the 1960s and Brazil and Colombia during the 1970s showed that economic growth can occur simultaneously with an increase in relative poverty within the population.[1] Bairoch (1975) provided evidence showing that although extensive economic growth in the developing countries occurred in the aftermath of World War II, the welfare of the populations in these countries did *not* improve. Lipton's study indicates a similar trend "... since 1945, growth and development in most poor countries have done little to raise the living standards of the poorest people" (Lipton 1977: 14). Lastly, Higgins comes to a similar conclusion:

> ... even countries with quite respectable rates of
> growth of national income did not enjoy
> development in the sense of wide and deep
> improvements in the welfare for the masses of
> the population.
>
> (quoted in Mandle 1980: 179)

At the same time, the experience of Pakistan, Nigeria and Iran during the 1960s and 1970s showed that the stresses and strains associated with economic growth can also have negative effects on the political climate in these countries, causing disruptions which in turn affect the economy.[2]

Thus, the growing realization among scholars and policy makers that the traditional GNP maximization efforts failed to produce substantial improvements in the lives of the low income population gave rise to a reordering of priorities which emphasized social factors. Success has been noted in South Korea, Taiwan, Thailand, and Yugoslavia in the achievement of the goal of "redistribution with growth," a World Bank slogan reflecting the effort to link increases in income per capita with a wider distribution of growth's benefits among the population. Recently, Lipton praised the success of Sri Lanka, Costa Rica and some states of India in their achievement of the multiple goals of development mentioned above: real economic growth occurred, public sector action was able to decrease poverty and these changes occurred in an environment of relative political freedom (Lipton 1984: 50). Many countries (for example, Sri Lanka and China) that did achieve success in the reduction of poverty experienced subsequent difficulties in sustaining the high degree of financial commitment by the public sector that an emphasis on social welfare demands. As a result of new information on the shortcomings of an approach addressing itself primarily to social factors, development efforts in the 1980s indicate a return to the growth objective in an effort to build the financial base that is necessary to finance the social projects.

In an effort to address issues such as the goals of development, this chapter examines the nature of the relationship between economic growth and the lives of the majority of the population in the states of India. Although it is not unreasonable to expect that as an economy grows, a general improvement in the welfare of the population will ensue, it does not necessarily follow that a high growth economy will provide for its people better than a low growth economy. This chapter presents some arguments pertaining to the relationship between growth and welfare improvements. It also tests empirically the hypothesis that there is a positive correlation between economic growth and an increase in welfare.

Lastly, the theoretical and empirical discussion is followed by attempts to identify the reasons for the unique performance in welfare indicators of one state, Kerala, by an examination of historical and political forces and trends.

Human welfare is a topic wrought with dispute and normative thinking, hence a discussion of it must commence with definitions. An increase in the general level of welfare consists of three component parts: (1) the satisfaction of basic needs, defined as the needs for "life-sustaining" goods such as food, health care, housing, etc., (2) employment opportunities, and (3) participation in the decision-making process. Although an analysis of political participation is beyond the scope of this book, it is noted that a relatively free political atmosphere exists in the states under study, given the democratic nature of India and the high degree of popular participation that is encouraged on all levels. The issue of the extent and nature of employment has already been addressed in chapters four and five. However, the concept of basic needs warrants elaboration.

"Basic needs" refers to an approach to development in which the focus is shifted from mere maximization of GNP to those needs which are most pressing and basic to human beings. There was a concensus among scholars (for example, Hicks 1979, Morawetz 1977, and Streeten 1979) that these needs include some agreed upon level of housing, food, health care, education, electricity, etc., although various scholars chose to extend the concept to include a healthy environment and popular participation (Blanchard 1977) or the right to a life with self-esteem (Goulet 1971).[3] Despite the innumerable difficulties that arise in determining how these needs can be satisfied, identifying the appropriate level of satisfaction and formulating measures of needs and their satisfaction, the basic needs approach was a unique effort at addressing development. Countries advocating this approach experienced impressive gains in such areas as literacy, nutritional levels, and life expectancy (for example, Cuba, China, and Sri Lanka during the 1970s). It emanated essentially from the research conducted at the World Bank in the late 1970s, during the period of the Bank's relatively humanistic orientation under the directorship of McNamara. It was during this time that the policy orientation of the Bank, as well as other donor agencies, shifted to an active involvement with problems of poverty, malnutrition and the like by redirecting their activities towards those in the bottom 40 percent of the income distribution. It must be noted that the chief proponent of the basic needs approach, the World Bank, has shifted away from this as the most important method of achieving social well-being. By the 1980s economists both in and out of the Bank began again emphasizing economic growth in recognition of the necessity of a strong economic base for the financing of all development efforts, including health, nutrition and education. The current free market, growth oriented approach of the World Bank and other donors does *not* negate the importance of the basic needs approach, but rather reflects added wisdom to development policy that occurred as a result of the basic needs experience.

Answers from Theory

Economic Growth Results in Welfare Improvement

The prevailing development theory and practice of the 1960s rested on the argument, rooted in neoclassical economic theory, that the goal of economic growth could be reached in the short run and poverty eliminated in the long run, under certain conditions. One of these conditions is the unequal distribution of income and assets within the country. The economic argument to justify inequality is as follows. The key to economic growth lies in the accumulation of capital. High rates of savings result in high rates of investment and high rates of capital accumulation. Given that high personal and corporate incomes are necessary to ensure savings, it follows that growth rates will be maximized if savings are maximized. If the rich have a greater marginal propensity to save and invest than the poor, who spend a significant portion of their income on consumption goods, then a greater concentration of income in the hands of the rich will result in a faster growing economy than one with a more equitable distribution of income.

The argument goes on to state that in the long run, poverty will be eliminated and even basic needs of the poor will be satisfied. This is the result of increased economic activity, generated by investments, which translates into increased demand for labor to enable increased production. Given a supply of labor to satisfy the demand, employment levels rise and consequently more people have incomes, enabling them to demand goods and services which satisfy their basic needs and consumptive desires. However, if income differences were to be completely eliminated, this trickle down process would not work, and therefore efforts at eliminating income inequalities would be self-defeating. Thus, one of the conditions for growth and the ultimate satisfaction of basic needs of the majority of the population is the persistence of inequality in the society. This strategy of development is safe insofar as it assumes the ultimate satisfaction of basic needs *without* actually having to address them in the short run. In the words of Bardhan (1984: 88), "For those who find radical institutional changes politically disturbing or unfeasible, the so called trickle-down effects of growth on poverty offer a comforting hypothesis."

<u>Welfare Improvements Result in Economic Growth</u>
Another argument describing the relationship between growth and welfare is based on causal linkages operating in the opposite direction from those described above. In other words, an investment in human resources (in the form of increased satisfaction of basic needs) results in economic growth. A situation of persisting poverty, in which a large portion of the population is underfed, underclothed and illiterate, tends to arrest growth because it contributes to low labor productivity. An emphasis on basic needs may enhance the future rate of economic development because a healthy, educated population becomes a skilled, adaptable labor force. In other words, addressing basic needs is *an investment in the population* which tends to result in long run productivity increases. Since productivity improvements are the basis of sustained, long term economic development (as described in chapter two), an investment in human beings results in greater growth, which in turn enables further investment in people, and thus a reinforcing and self-perpetuating process occurs.

A complication seemingly arises because basic needs services in most countries are provided by the public sector. Even when the overall economy is based on private enterprise, the public sector is often the chief provider of educational, health care, housing and electrical services for a large segment of the population. This is a result of low profits and inefficiency associated with their production. To increase the provision of those services, the public sector must siphon funds (in the form of taxation) from those with surplus income, thus possibly reducing their savings. This may result in a decrease in investment, as well as the rate of economic growth and consequently employment. However, it may be argued that this is only a short run phenomenon and would not persist in the long run as the increased productivity of the labor force tends to entice producers to hire more labor relative to other inputs, and thus rectify the unemployment problem.

Another way in which a strategy aimed at the bottom 40 percent of the population will stimulate economic growth is by the increase in demand for consumer goods that will result from an increased income among this population group. Given the evidence from many developing countries that the rich tend to save and consume imported luxury articles whereas the low income population has a higher marginal propensity to consume, an increased income among the poor will in all likelihood be spent on domestically produced articles, and thus would stimulate local production, local employment and local investment.

On the basis of increased labor productivity and demand generation, scholars advocating a development strategy based on investment in human resources claim that on balance, increasing incomes of the poor tends to be growth promoting rather than growth retarding (for example, Hicks 1979).

Answers from Practice

It was noted above that various studies of developing countries where high rates of growth occurred showed no, or at best, insignificant improvements in the welfare of the majority. The most comprehensive global study indicating this phenomenon was undertaken by Morawetz (1977). It entailed the correlation of economic growth and various indicators of the satisfaction of basic needs (specifically nutrition, health, housing and education), in some sixty countries. Statistical analysis indicated the lack of a strong correlation between economic growth and welfare improvements. Would similar results be observed within the agricultural states of India? Has the economic growth which occurred in the agricultural states of India been associated with an improvement in the lives of the Indian population, where improvement is measured by the increased satisfaction of basic needs? Does "more growth" imply "more welfare?" Empirical evidence from various states in India, for the period 1961-1981, is presented to answer these questions.

Empirical Evidence from the Agricultural States of India
(1) Economic Growth in Indian States

Based on the data presented in chapter two, there is no doubt that all the agricultural states of India experienced an increase in their incomes per capita during the 1960s and 1970s. However, Punjab and Haryana grew at the highest rates (3.1 and 2.9 percent average annual growth, respectively, during the 1960s), and attained the highest levels of income per capita by 1971 (and retained this position in 1981: 1380 and 1061 constant Rs., compared to the Indian average of 697). A.P. and Bihar experienced the lowest rates of growth during this period. Given then that economic growth occurred, to what extent did this process result in an improvement in the welfare of the majority?

(2) Basic Needs in Indian States

The levels and changes over time of several indicators of basic needs were observed, and these were then correlated with the corresponding state growth rates. The indicators were chosen with Morawetz's study as a guideline, but due to data constraints, not all were available for all states over all years under study. Consequently, the three basic needs indicators observed were education (measured by literacy rates and number of schools per population), health care (measured by infant mortality rates and number of beds in hospitals and dispensaries), and electricity (measured by the percentage of villages electrified, and domestic fuel consumption). Whenever possible, the choice of indicators reflects active participation rather than mere availability, since satisfaction of basic needs implies that services made available were actually consumed.[4] These three are only approximate indicators of welfare since an important component of basic needs, i.e., the need for food, was omitted as a result of unsatisfactory regional data.

Health

It is evident from Tables 6.1 and 6.2 that there has been an unequivocal improvement in health conditions in nine of the ten states.[5] The infant mortality rates have decreased during the 1960s everywhere except in M.P., where a slight increase was noted. Similarly, the number of beds available for medical treatment increased in all states except for M.P., where there was no change from 1959 to 1973. These indicators of health care services suggest improvements in welfare irrespective of economic growth, since the value of the health indicators do not correlate significantly with economic growth (the growth rates are presented in chapter two, Table 2.1). In fact, when growth values are regressed upon the value of infant mortality rates in 1973 and 1978, the correlation coefficients are .27 and .11 respectively. Similarly, the coefficients in the case of beds available for medical treatment are .32 and -.12. In other words, although Punjab and Haryana, the high growth states, have experienced increases in the number of beds available and a decrease in the infant mortality rates, these changes are by no means greater than in the states which experienced lower levels of economic growth. Kerala, a moderate growth state, had by far the lowest infant mortality rate in all

TABLE 6.1

Infant mortality rates (per 1000 live births)

	1960	1973	%change (1960-73)	1978	%change (1973-78)
Punjab	86.1	54.0a	-37.2	46.0	-14.8
Haryana	57.2	48.0b	-16.1	48.0	0.0
Karnat	61.6	47.0	-23.7	39.0	-17.0
U.P.	91.7	59.0	-35.6	43.0	-27.1
Orissa	135.0	69.0	-48.9	62.0	10.2
M.P.	93.0	95.0	2.2	67.0	-29.5
Kerala	39.8	18.0	-54.8	17.0	-5.5
T.N.	91.6	49.0	-46.5	40.0	-18.4
A.P.	79.5	45.0	-43.4	35.0	-22.2
Bihar	80.4	33.0	-59.0	31.0	-6.1
r^2		0.27		0.11	

a-1968,
b-1966,
Source: various tables in various yearly volumes of Government of India, Ministry of Home Affairs, *Vital Statistics of India*, New Delhi.

TABLE 6.2

Hospital and dispensary beds (per 100,000 people)

	1959	1973	%change (1959-73)	1978	%change (1973-78)
Punjab	71.2c	73.5	3.2	89	21.1
Haryana	48.8c	70.0	43.4	62	-11.4
Karnat	56.0	95.6	70.7	86	-10.4
U.P.	40.7	45.3	11.3	47	3.8
Orissa	20.1	50.0	148.8	45	-0.1
M.P.	40.6	40.8	0.5	34	-16.7
Kerala	75.8	107.5	41.8	196	82.3
T.N.	62.9c	68.0	8.1	95	39.7
A.P.	53.8c	71.3	32.5	67	-6.0
Bihar	15.3	28.4	85.6	34	19.7
r^2		0.32		-0.12	

c-1961
Source: various tables in various yearly volumes of *State Statistical Abstracts*.

three years under study, as well as the largest number of beds per population. It therefore cannot be concluded that high rates of growth are associated with a disproportionately large improvement in public health.

Electricity

Tables 6.3 and 6.4 indicate that from 1961-62 to 1972-73 domestic fuel consumption increased in all states, although most dramatically in Karnataka, a moderate growth state. The correlation coefficient of growth regressed upon the value of fuel consumption in 1972-73 was .57. There was an increase in the villages and towns electrified over the course of 1961 through 1983, resulting in 100 percent electrification in Punjab, Haryana and Kerala. The results of a regression analysis yielded correlation coefficients of .56 and .38 between growth and the level of electrification in 1975 and 1983 respectively. This evidence suggests that high rates of economic growth do not necessarily imply a commensurately large increase in electrical capacity.

Education

Literacy rates, as measured by the percent of the population that is literate, increased in all states during the 1960s and the 1970s. During the first decade, these increases vary between 23 and 35 percent except in the case of Bihar and A.P., where the literacy rates changed by a mere 8.4 and 15.9 percent. Kerala and T.N. both had final literacy values which exceeded those observed in the high growth states. In the second decade, the percent change ranged from 16.6 to 34.7. It must be noted that both the initial and final literacy rates in Kerala were higher than for the remaining states (46.9% in 1961 and 70.4% in 1981, respectively). Regression analysis showed that the value of literacy achieved in 1971 and 1981 regressed upon growth yielded .07 and -.09 respectively.

The number of primary schools per 10,000 people increased significantly only in U.P., whereas it remained virtually constant in five states and decreased in four, two of which are the high growth states. However, during the 1970s, Punjab and M.P. experienced increases. The correlation coefficients of growth regressed upon the values of primary schools are .4 in 1970-71 and -.11 in 1976-77. Both of these indicate that high rates of growth do not imply a greater increase in the availability of primary education among the Indian population than would otherwise be the case.

TABLE 6.3

Domestic fuel consumption (in kwh per capita)

	1961-2	1972-3	% change
Punjab	8.7a	12.1	39.1
Haryana	4.0b	7.6	90.0
Karnat	3.2	9.9	209.4
U.P.	1.7	4.5	164.7
Orissa	1.2	1.4	16.7
M.P.	1.4	3.7	164.3
Kerala	3.2	5.9	84.4
T.N.	4.9	8.8	79.6
A.P.	2.2	4.4	100.0
Bihar	1.1	1.9	72.7
r^2		0.57	

a: 1967-68,
b: 1965-66
* includes Union Territories
Source: various tables in various yearly volumes of Statistics Division, *Fertilizer Statistics*, Fertilizer Association of India, New Delhi.

TABLE 6.4

Towns and villages electrified (as percent of total)

	1961	1975	%change (1961-75)	1983	%change (1975-83)
Punjab	17.1	64.6	227.8	100.0	54.8
Haryana	8.5	94.4c	1010.6	100.0	5.9
Karnat	10.5	52.1*	396.1	67.9	30.3
U.P.	4.3	27.4	537.2	43.2d	57.7
Orissa	.25	21.8	8620.0	45.3	107.8
M.P.	.53	16.1	2937.7	46.3	187.6
Kerala	55.3	75.1	35.8	100.0	33.2
T.N.	17.8	41.5*	133.2	99.4	139.5
A.P.	9.0	38.6	329.3	75.9	96.6
Bihar	3.3	15.0	354.6	43.2	188.0
r^2		0.56		0.38	

c: 1972-73,
d: 1982
* includes Union Territories
Source: see Table 6.3

TABLE 6.5

Literacy rates

	1961	1971	%change (1961-71)	1981	%change (1971-81)
Punjab	26.7	33.7	26.1	40.9	21.4
Haryana	19.1	26.8	35.1	36.1	34.7
Karnat	25.4	31.5	24.1	38.5	22.2
U.P.	17.6	21.7	23.0	27.2	25.3
Orissa	21.7	26.2	20.9	34.2	30.5
M.P.	17.1	22.2	29.3	27.9	25.7
Kerala	46.9	60.4	29.0	70.4	16.6
T.N.	31.4	39.5	25.6	46.8	18.5
A.P.	21.2	24.6	15.9	29.9	21.5
Bihar	18.4	19.9	8.4	26.2	31.7
r^2			0.07		-0.09

Source: various *State Statistical Abstracts*.

TABLE 6.6

Primary schools (per 10,000 people)

	1960-61	1970-71	%change (1961-71)	1976-77	%change (1971-77)
Punjab	6.3a	5.4	-14.3	6	11.1
Haryana	5.9b	4.2	-28.8	4	-4.8
Karnat	8.9	7.4	-16.8	6	-18.9
U.P.	5.4	7.0	29.6	6	-14.3
Orissa	12.5	12.6	0.8	12	-4.8
M.P.	8.6	8.9	3.5	10	12.4
Kerala	12.5	3.2	-74.4	3	-6.3
T.N.	7.0	6.3	-10.0	6	-4.8
A.P.	9.5	8.5	-10.5	7	-17.7
Bihar	8.0	8.3	3.8	7	-15.7
r^2			0.4		-0.11

a: 1966-67
b: 1965-66
Source: various *State Statistical Abstracts*.

Economic Growth and Basic Needs

On the basis of a time series analysis, it was found that, over time and in most states, improvements occurred in most basic needs indicators. However, the cross sectional analysis showed that these improvements were not related to high rates of economic growth, since the correlation coefficients of growth regressed upon various indicators of basic needs never exceeded .57. Thus, the validity of Morawetz's global conclusions pertaining to the lack of correlation between high growth rates and basic needs satisfaction was confirmed in the particular context of the ten Indian states. Empirical evidence shows that the high growth states of Punjab and Haryana did not distinguish themselves from the remaining eight states in the provision of welfare to their people. The growth that was achieved did not, in these states, translate into a greater improvement in the lives of the inhabitants than occurred elsewhere. It is in fact Kerala, a medium growth state, that surpassed all others by most indicators of basic needs. During the 1960s and 1970s, it had by far the lowest infant mortality rates, highest number of beds per population, highest literacy rates and number of schools. It also had the second highest percentage of villages electrified. This impressive performance in most indicators of basic needs points to the uniqueness of Kerala.

Discussion

Given that all states, high or low growth, experienced some degree of improvement in welfare, several questions arise: why wasn't the economic growth in India associated with greater improvements in the satisfaction of basic needs? Why weren't the levels of the indicators of basic needs in the high growth states higher than in the low growth states and why did they fail to increase at a greater rate?

Let us recall that in developing countries, the public sector tends to provide social services aimed at improving basic needs, and India is no exception. Given the decentralized nature of the Indian Union, state governments, rather than the central government, provide these services.[6] It is the goals and objectives of individual state governments that affect the nature of the growth process occurring within the states since the public sector has the role of directing and guiding the development process in accordance with its goals. The degree of commitment to eradication of poverty might be identified by observing the per capita state expenditures on social services. A review of this expenditure in the Indian states

TABLE 6.7

Per capita government expenditure on education, in Rs.

State	1965-66	1969-70	1976-77
A.P.	7.67	12.22	28.01
Bihar	3.89	8.65	13.81
Haryana	na	16.41	35.26
Kerala	15.19	25.18	59.45
Karnataka	9.53	15.28	32.32
M.P.	8.49	11.66	22.68
Orissa	5.83	9.73	26.86
Punjab*	9.63	18.97	47.12
T.N.	10.77	16.09	31.55
U.P.	5.68	7.87	21.64

*1965-66 data refer to Erstwhile Punjab
Source (first column): Government of Kerala, (1966) *Kerala: An Economic Review*,
Bureau of Economics and Statistics, Trivandrum, Tables 9.2
(second column): ibid, Table 10,
(third column): Government of Kerala (1979), *State Handbook of Kerala*, Bureau of
Economics and Statistics, Trivandrum, Table 33.3.

TABLE 6.8

Per capita government expenditure on health services, in Rs.

State	1965-66	1969-70	1976-77
A.P.	3.21	5.27	13.88
Bihar	2.09	3.05	6.74
Haryana	na	5.80	15.59
Kerala	4.49	6.96	18.35
Karnataka	3.03	4.95	15.57
M.P.	2.89	4.97	13.30
Orissa	2.94	4.96	11.79
Punjab*	3.64	6.06	18.48
T.N.	3.71	5.90	17.10
U.P.	2.07	3.57	7.84

*1965-66 data refer to Erstwhile Punjab
Source: see Table 6.7

(Tables 6.7 and 6.8) indicates that the public sector in Kerala incurred greater per capita expenses for education and health care than any other state, including the high growth states, during the years 1965 through 1977.

The extent to which economic development is accompanied by an improvement in welfare depends partially upon the degree of public sector expenditures, which in turn depends upon: (i) the commitment of the state government to attain this goal, and (2) the government's ability to do so.[7] The goals and expenditures of the state governments are described below, and the case of Kerala is compared to that of the high growth states. Although it is recognized that in such a comparison the *ceteris paribus* condition does not apply, some hypotheses are nonetheless offered to explain the differences in basic needs satisfactions.

The Capacity to Attain Welfare Goals: Kerala and Punjab/Haryana

A government's capacity to improve the collective welfare is largely dependent upon revenues which can be tapped for social services. Among possible sources of revenue, those generated internally are the most important in the Indian states. Despite the problems associated with agricultural taxation, the greater the economic growth, the greater the taxable income to be allocated for the attainment of social goals. There is no doubt that more income was generated in high growth states during the green revolution than in either Kerala or the remaining states, hence more income was potentially available for appropriation through fiscal measures. However, it is necessary to stress the *potential* aspect of the government's capacity to redistribute income because it has been found that the actual taxes paid by the rural population are minimal or nonexistent as a result of the rural power structure which favors the wealthy farmers. Nonetheless, the capacity to address basic needs was greater in Punjab and Haryana than in Kerala, but was this also true of the incentive?

The Incentive to Attain Welfare Goals: Kerala

State governments often proclaim the necessity and desirability of economic growth and eradication of poverty. The improvement in welfare is sometimes a sincere goal and sometimes it is merely paid lip service. In Kerala, the social wellbeing of the population is among the government's primary concerns. The experience of Kerala is unique and warrants elaboration. There exist at least two possible reasons why the Kerala governments might have been motivated to expend large portions of their budgets on social services. One reason, greatly disputed in the literature, focuses on the social ideology of the leftist governments that have, at various times since Independence, ruled Kerala. The second reason stems from the historical influence of pre-Independence rulers of the regions. Both will be reviewed below.

(1) The Social Policy of the Leftist Governments in Kerala

The left was particularly strong in Kerala politics before and after Independence.[8] A communist government was voted into power in 1957 in Kerala, and in 1959 removed from office by the central government. There followed a period of rule by Presidency (from the Center) and a series of Congress governments which coexisted with a large number of leftist parties in the Assembly. Finally, ten years after the communists were removed from power, the United Front, consisting of a union of leftist parties under the guidance of the communists, returned to head the government for two years. Although the communists led the state government for a relatively short time, the presence of various leftist parties has been a fixture in Kerala's political scene since before Independence.[9]

Is the relatively high expenditure on social services in Kerala related to the particular social ideology of Kerala's leftist governments? Horvat (1979) and Mandle (1980) studied the relationship between basic needs and economic growth in socialist and non-socialist countries and concluded that in the socialist countries, such as Yugoslavia, Hungary and Cuba, basic needs seemed to be more satisfied.[10] Thus we might expect greater social service allocations from leftist governments in Kerala. The literature on the subject is varied. T.J. Nossiter (1983) points out that the communists in Kerala contributed substantially in the advancement of social welfare. Although most achievements of the leftist governments are in the realm of land reform and industrial and labor policy, Balaram (1973) claims that with respect to welfare schemes, the heavy annual expenditure on education resulted in "big strides" in literacy and primary education.[11] Sukumaran Nayar (1969) also claims that the communist government of 1957-1959 was effective in its social policy, although it was restrained from doing more in promoting economic development and social welfare by the Congress rule in New Delhi.

A review of Kerala's budgets for health and education indicates that there is little difference between the various governments. Table 6.9 shows that although the absolute budget allocation to education and health grew during the mid-1950s through the mid-1970s, the percentage of total development expenditure allocated to these social programs changed significantly only in the early 1960s.

A comparison of the expenditure data with the rule of leftist governments shows that the communist party's social policy did not result in greater budget allocations for education and health. In fact, the largest percentage increase in expenditure coincided with a Congress government. Overall, there were no major differences in social expenditure between the leftist and Congress governments. This observation agrees with Nayar's claim that "there is certainly no empirical evidence to show that one political elite (in Kerala) was better than another in effecting planned social change" (Nayar 1972: 273).

TABLE 6.9

State budget allocated to education and public health in Kerala (in Rs. x 100,000)

	(1956-57)	(1961-62)	(1966-67)	(1971-72)	(1972-73)
actual	1253	2391	4466	8481	8990
%	54.9	54.5	63.9	65.1	62.3

note: % refers to percentage of total development expenditure.
Source: calculated from Government of Kerala (1975), *Statistics Relating to the Kerala Economy 1956-7 to 1973-4*, Bureau of Economics and Statistics, Trivandrum, Table 14.2.

(2) Historical Influences

Given that neither economic growth nor the social policy of the leftist governments are directly responsible for the high basic needs indicators in Kerala, we turn to history in order to identify the source of its uniqueness. Modern Kerala differs from the rest of India in part as a result of historical circumstances which influenced its social and economic development. Today, Kerala consists of former Travancore, Cochin and part of Malabar. The pre-Independence princes of these regions, and to a lesser degree those of neighboring Baroda and Mysore, placed great emphasis on the social condition of their populations. According to Sreedhara Menon (1978), as early as the 13th century, under the rule of Ravi Varma Kulasekhara, parts of Kerala became centers of learning and culture. The "royal patronage of learning," so unique to south Asia, reached new dimensions in the 18th century with the introduction of the first printing press in India.[12] However, the concerns of the princes of Travancore and Cochin transcended learning and culture. Historians claim that they emphasized social services to a greater degree than their counterparts throughout the subcontinent. Among these, Sukumaran Nayar (1969) says that "the economic and social policy in [these] princely states was more progressive than in British India", whereas Menon (1969) claims that due to "the enlightened Maharajas and their capable ministers [in Kerala], social reforms made headways while in British Indian and [in other] princely states they lagged".[13]

Perhaps awareness of the social condition and openness to social reform that prevailed in south India produced a receptive atmosphere which accommodated and encouraged missionary work, especially in the areas of education and health. There is evidence of Catholic Portuguese missionary activity in the 18th century in Travancore, Coimbator and the Malabar coast. In 1813, the British withdrew their opposition to missionary activity and this date marks a new surge of American, German and other missionary penetration into south India. Many scholars studying this period of Indian history agree that the missionary activities were primarily educational: for example, Wolper (1977) points out that missionaries introduced a concern for widespread education of all segments of the population which resulted in a large network of colleges throughout the region.[14] There is evidence that Christian missionaries in the early 1800s believed they could eliminate the existing social inequalities resulting from the Hindu caste system through education. Schools and colleges were opened first for Christian males, then for females and members of the backward castes. This concern with education was shared by the Travancore government which, by 1904, bore the entire cost of education of lower caste children (Yesudas 1980: 201). Missionary activity, however, was not limited to education. According to Spear (1961), missionaries organized and provided community health services which "touched all classes", thereby extending their influence beyond the lower castes.[15] Their activities in the hospitals and dispensaries, coupled with the efforts of the princely rulers, resulted in health care provision on a scale unprecedented in the history of south Asia.

The extent of missionary activity and influence in Kerala is evident by both the present size of the Christian population as well as the pre-Independence literacy rates. Specifically, one might measure the effect of missionary activity by the conversion to Christianity that occurred: in 1971, 31.6 percent of the Indian Christians reside in Kerala, amounting to

22 percent of Kerala's population.[16] However, many took advantage of the schools and health services provided by the missionaries without necessarily accepting Christianity, as evident by Kerala's superior performance at the time of Independence in at least one indicator of basic needs, literacy (mortality rates from this period are too unreliable to be presented). Specifically, the literacy rates in Kerala in 1951 were much higher than in the remaining states: 40.7 percent of the total population was literate, followed by 24.0 percent in West Bengal and 15.2 percent in Erstwhile Punjab. This indicates that the post-Independence governments of Kerala inherited a relatively well educated population.

It is then suggested that the influence of an historical pattern of undisputed focus on social welfare issues in Kerala transcends relatively high literacy and low mortality rates, but more importantly, leaves its mark on the relationship between the population and its rulers. Given that "the descendants of the 19th century population are well educated and politically conscious" (Nyrop 1975),[17] *this educated electorate might have exerted pressure on the government to address social matters* in a manner and to a degree that failed to occur elsewhere in the country. In fact, we found that whether the ruling party was Congress, United Front or CPI, the ideology with respect to social services did not change significantly.

Thus, it seems that the particular historical circumstances set the stage for the relatively higher basic needs satisfaction in Kerala than elsewhere. The Kerala governments of the 1960s and 1970s had an incentive to address social welfare issues, and to incorporate an improvement in welfare into the process of economic growth in a way that other states lacked.

The Incentive to Attain Welfare Goals: Punjab/Haryana

The high growth regions did not have the historical experience of Kerala, hence lacked the popular pressure for emphasis on social services. In addition, the governments of the high growth states lacked the incentive to take strong action in the provision of basic needs services for another reason, described below. To attain welfare goals such as the increased satisfaction of basic needs, the public sector must in effect redistribute income. Although the governments of Punjab and Haryana had the legal authority to do this, their growth experience was based on inequality, and might not have sustained itself under conditions of equality resulting from a serious redistribution attempt. During the green revolution, income inequalities[18] within the region intensified since the adoption of the new technologies was more prevalent among the wealthy, large farmers who were capable of mechanizing, who were more likely to secure credit, who were able to increase their production and thus take advantage of the profitability of the new seeds. Increased concentration of capital in the hands of some farmers enabled greater production and accumulation of profits. A large scale interference in these profits by the government would greatly decrease their working capital, hence their production, savings, and capital accumulation. Therefore, so as not to interfere with this process, the basic needs of the majority were to be satisfied by the increased employment and income that would trickle down as a result of economic growth. In other words, the neoclassical argument that growth ultimately results in welfare improvements was the basis of the welfare policies of Punjab and Haryana in the recent decades. Thus, although the high growth regions had more taxable income, their governments had less incentive to alter the nature of the agricultural growth process in order to attain of welfare goals.

Conclusion

The empirical evidence from the agricultural states of India indicates that the attainment of basic needs goals was, in the aggregate, no more successful in high growth regions than in low growth regions. In the context of a developing economy, where the public sector plays an important role, the governments must have the incentive as well as the capacity to achieve increased basic needs satisfaction. In Kerala, the incentive existed, although the capacity was lacking. In the high growth states, Punjab and Haryana, the capacity was evident, but incentives were weak. Thus, in all the states, either the ability or commitment (or both) was lacking and economic growth failed to improve the lives of the majority as it might have under different conditions.

Notes

1. Montek Ahluwalia (1973), "Dimensions of the Problem" in Chenery, Duloy and Jolly (eds.), *Redistribution with Growth: An Approach to Policy*, Washington: IBRD mimeo.

2. Albert Hirshman (1981), "The Rise and Fall of Development Economics" in *Essays in Trespassing: Economics to Politics and Beyond*, New York: Cambridge University Press.

3. F. Blanchard (1977), *Employment, Growth and Basic Needs: A One-World Problem*, New York: ILO, and Dennis Goulet (1971), *The Cruel Choice*, New York: Atheneum.

4. These indicators were adopted despite their criticism by Hella (1983, "Basic Needs and Economic Systems: Notes on Data, Methodology and Interpretation" in *Review of Social Economy* 41/2) because they represent the closest indication of the consumption of basic needs services.

5. In Tables 6.1 through 6.6, states are ranked according to the growth they experienced during the 1960s.

6. In fact, in India, basic needs services such as education, health care, and electricity are essentially in the domain of the public sector. There is some private provision of some of these services, but this rarely affects the majority of the population.

7. Given the nature of center-state relations in India, one is justified in observing state level social policy and decision making. Indian states have a large degree of autonomy with respect to budgeting and resource allocation (see S.A.H. Haqqi (ed.), (1976), *Union-State Relations in India*, Meerut: Meenakshi Prakashan).The central plan sets the guidelines which leave state governments ample independence, especially with respect to social services. In fact, M. Adiseshiah (1973, "Education in Center-State Relations" in B.L. Maheshwari (ed.), *Center-State Relations in the 1970s*, Calcutta: Minerva Association) points out that the constitution makes education a state responsibility (and hence the wide state differences in quantity and quality).

8. The varied political past of Kerala was addressed by many scholars, among which are Fic (1970), Nayar (1972), Nossiter (1983), Sukumaran Nayar (1969).

9. The presence of leftist movements was felt in this region long before Independence. E.M.S. Nambooriripad (in Sukumaran Nayar 1969) discusses peasant conflicts which resulted in political struggles for change. These include the Malabar rebellion of 1921, the Non-cooperation movement of 1921-22 and the *Saltsatyagaraha* of 1932.

10. Branko Horvat (1974), "The Welfare of the Common Man in Various Countries" in *World Development* 11/7.

11. N.E. Balaram (1973), *Kerala: Three Years of U.F. Government*, Communist Party Publication #10, New Delhi: New Age Printing Press, p.43.

12. A. Sreedhara Menon (1978), *Cultural Heritage of Kerala*, Cochin: East-West Pub., p. 170.

13. V.K. Sukumaran Nayar (1969), "Political Aspects of Economic Development in Kerala" in V.K. Sukumaran Nayar (ed.), *Kerala Society and Politics*, Trivandrum: Indian Political Science Conference, p. 262, and A. Sreedhara Menon (1969), "Kerala History" in V.K. Sukumaran Nayar (ed.), *Kerala Society and Politics*, ibid., p.14.

14. S. Wolpert (1977), *A New History of India*, New York: Oxford University Press, p. 241.

15. Percival Spear (1961), *India: A Modern History*, Ann Arbor: University of Michigan Press, p. 290.

16. These data are taken from Nyrop et al. (1975, *Area Handbook for India*, Washington: U.S. Government Printing Office), Table 1.9. The Christians in Kerala are second only to the Hindus, and are more numerous than the Muslims.

17. Nyrop et al. (1975), ibid.

18. This effect is not limited to the Punjab and Haryana. Among the extensive literature on inequality and the green revolution, see, for example, Frankel (1971), Griffin (1974), and others mentioned in chapter two.

7

Interstate Migration in India

Population movements have occurred throughout history. These migrations have differed with respect to destination and origin, source of motivation ("push" vs. "pull"), and characteristics of the migrant. For example, the transcontinental large scale movements of workers to the "new world" (Argentina, Australia, Brazil, North America) have been replaced by interregional and often temporary migrations. In addition, whereas 18th and 19th century migrations were characterized by population movements from rich countries to those that were generally less developed, recent population flows originate in regions that tend to be poor and overpopulated and are destined for the more developed regions. Various regions of the world are characterized by an unskilled population exodus (e.g., from African countries to the oil-abundant Middle East), whereas others suffer from a severe brain drain (e.g., India).

Empirical evidence from numerous countries suggests a strong association between migration and economic growth. The pioneering work of Lee, Kuznets and Eldridge (1964)[1] on interstate migration patterns in the United States points out the dual role of migration in the process of economic growth: it not only contributes to growth but is also one of its consequences.[2] A region undergoing economic growth tends to attract population by virtue of the economic opportunity it offers, thus exerting a pull on the migrant. This opportunity reflects the changing economic structure associated with economic growth. Specifically, new consumption patterns (resulting from economic growth) stimulate structural changes in production, which in turn generate a new pattern of demand for labor. Since industries tend to locate in urbanized centers while the pre-industrial production tends to be rural and based on agriculture, an excess supply of labor develops in rural areas (contributing to the push of population), while excess demand for labor emerges in urban areas.[3] Furthermore, the potential of the natural population increase to satisfy the largely urban labor needs is outweighed by the magnitude of the industrial shifts. All these tendencies contribute to migratory flows of people by which *the changing manpower requirements of the growing economy are met.* Migration patterns are thus closely related to the spatial distribution of economic activity.[4]

In addition, the characteristics of the migrant affect economic growth in both gaining and losing regions.[5] Economic growth accentuates the differences among the population and causes the more capable, brighter and enterprising individuals to rise to the surface. It is these "positive" individuals that are more likely to succeed and often the ones more likely to take the courageous step of migrating in order to avail themselves of economic opportunity elsewhere. Thus, outmigration may result in the loss of positive members of the population and may thus decelerate a region's pace of growth. At the same time, the addition of capable individuals to the labor force will accelerate the receiving region's growth rate. Economic growth may therefore exert both a pull and push on the migrant, so that the migration pattern is the net result of these offsetting tendencies.

This chapter contains an analysis of migration in the ten agricultural states of India. In addition to the presentation of migration estimates, some possible influences on interstate migration are discussed, such as the source of growth, population density and economic growth both in agricultural and industrial regions. The relationship between migration and economic growth is studied in an effort to assess whether the determinants of U.S. migration, as discussed by Lee, Kuznets and Eldridge (1964), apply to India. *State level* migration patterns are observed, and a breakdown by district is not included. Although state level movements obliterate small scale and short term migration often associated with seasonal changes in demand of labor, they do encompass long term, long distance trends based on a greater degree of permanence. In order to classify states as absorbing or losing, only *net* population movements are observed.

How does an analysis of migration fit into this study of agricultural India? The theoretical framework presented above is relevant here insofar as the agricultural sector tends to lose population to regions where industrialization and services are concentrated. Therefore, population migrations are related to various aspects of economic development, including fluctuating labor demands associated with both the changing importance of sectors as well as innovative technology (chapters two and three). Indeed, most issues discussed in this book are related to migration: the ability of females to respond to labor demands that entail migration (chapter four), the location of industry and its role in the migratory process (chapter five), as well as the role of social services in the creation of an environment conducive to satisfying and thereby "containing" its residents (chapters six and eight).

Empirical Evidence of Migration in India

Background

To students of development, India is an anomaly with respect to its migration patterns. High degrees of internal migration, such as those experienced in Latin America, were often predicted yet never materialized. Prior to 1930, the Indian population was extremely immobile, although British and Princely rule never formally or legally restricted labor mobility (Davis 1951).[6] The early post-World War II censuses stimulated several studies of internal migration patterns (Gosal and Krishnan 1975, Mitra 1968, SenGupta 1968, Zachariah 1964), all of which indicate the increase in internal migration since Independence and Partition. A few statistics will illustrate this point:[7] in 1921, 9.8 percent of the total population (or 24.7 million people) was enumerated outside their district of birth, increasing to 10.8 percent in 1951 (38.5 million people) and 12.1 percent in 1961 (52.9 million people). The 1961 census furthermore indicated that one half of the total migrants to date moved during 1951-1961 alone. This general increase in population shifts was mostly rural-rural (73.7% of migrants), followed by rural-urban (14.5%), urban-urban (8.0%) and lastly urban-rural (3.6%). The rare state level migration estimates show that, in 1961, 3.3 percent of the population was enumerated in a state outside the state of birth, resulting in the following pattern among the agricultural states: Karnataka was the only receiving state in 1951, whereas by 1961 M.P. also had net inmigration (Vaidyanathan 1967: 67). In sum, India has experienced a *steady increase in population movements*, most of which have been attributed to perceived economic opportunity or family and marriage reasons (applicable mostly to female migrants).

Another anomaly of Indian migratory patterns was pointed out by Becker, Mills and Williamson (1986) in their discussion of the relationship between migration and the manufacturing sector. Specifically, they showed that although employment in the organized urban sector stagnated during the 1960s and 1970s, the urban population grew twice as fast as the rural population, largely as a result of migration from the rural regions. This was found to be inconsistent with existing theory which predicts migratory movements in response to perceived economic opportunity.

Lastly, it is noted that international migration is insignificant in *magnitude* in the Indian context. It is estimated that only 0.7 to 0.8 percent of the population resides abroad (amounting to approximately 5 million people). However, the *impact* of international migration is significant because of the characteristics of the migrants: according to Madhavan (1985), since the Second World War, international migrants have been predominantly highly skilled workers, thus differing from the prewar period when most emigrants were the rural poor. Therefore, although small in size in comparison to the total population and total migration, emigration does have a long term impact on the Indian economy.

Migration during 1961-1971

Premi's (1982) estimates of migration constitute the most comprehensive review of population movements during the 1960s. However, only the total net interstate migration rates are presented in Table 7.1. In addition, another set of observations, estimated by the National Research Council (1984), under the guidance of Bhat, Preston, and Dyson is included. Two sets of observations are included because of discrepancies arising from different methods of estimation. Although the direction of net migration coincides in both studies, the magnitude of the rates differs. For example, Punjab and U.P. differ in their ranking as the most outmigrating states. In addition, there is disagreement between the ranking of Orissa and Haryana, as well as T.N. and A.P. (although in the latter case, the magnitudes of difference are insignificant).

S.K. Sinha (1975) also grouped states according to their migration rates, and found Punjab to be the greatest loser of population. He also identified an insignificant difference between Haryana and Orissa, and A.P. and T.N., thereby aligning his estimates more closely with Premi's.

Thus, although different methods yield different migration results, the trends observed by three scholars seem to indicate the following: during the course of the 1960s, Punjab, U.P., Kerala and to a lesser degree Bihar were net losers of population, M.P., Orissa and Haryana absorbed more people than they lost, whereas in A.P., T.N. and Karnataka there were insignificant net population movements.

TABLE 7.1

Average yearly net migration rates (per 1000), 1961-1971

State	Premi	NRC
A.P.	-0.32	-0.21
Bihar	-1.00	-0.86
Haryana	0.74	0.65
Karnataka	0.31	0.18
Kerala	-1.56	-1.23
M.P.	0.83	0.78
Orissa	0.32	0.66
Punjab	-1.87	-1.27
T.N.	-0.30	-0.34
U.P.	-1.28	-1.30

Source: computed from Premi, 1982: 142 and National Research Council (NRC), 1984: 112-113.

Other information conveyed by Premi's estimates sheds light on the migratory patterns in India during the 1960s.[8] First, the trend in migration is identified. For example, Punjab has experienced a slowdown in its migration rate since only 38.4 percent of its migrants relocated during the intercensal period. Individuals that migrated in the last year of the study, i.e. 1970-71, accounted for only 1.5 percent of the total population movement. In the remaining states, intercensal migration as a percentage of total migration ranges from 43 to 56 percent, indicating that approximately half of the migrants changed residence during the decade under study and implying that interstate migration either increased or remained constant over time in those states.

Second, Premi's estimates indicate state differences with respect to population movements by sex. In all states under study except Haryana and Orissa, there have been more male migrants, over time, than female. In Haryana, there were more females inmigrating than males, both in absolute terms (36,000 females compared to 19,000 males) and as a percentage of total population (0.77 and 0.35 respectively). In Orissa, there was a net outflow of 15,000 males, and a net inflow of 49,000 females (as a percentage of the population, 0.14 males and 0.45 females are migrants).

Third, it is noted that total migrants as a percent of the population are the greatest in Punjab (4.88 percent of the population), followed by Kerala (2.77), U.P. (2.44) and Bihar (2.32). In the remaining states, migrants are an insignificant component of the population.

Lastly, with respect to the role of migration in population changes among states, India has historically differed from the USA or Europe insofar as its high rates of natural population increase far outweighed migration in accounting for differing rates of population growth. According to Premi's estimates, intercensal migrants as a percentage of intercensal population growth vary greatly among states. It is only states in which outflows of population are great (Punjab, U.P. and Kerala) that migration accounts for a significant part of intercensal population change (in these states, intercensal migrants as percentage of intercensal population growth are 19.0, 13.7 and 13.4, respectively; Zarkovic 1984: 87).

Migration during 1970-1981

The migration rates for the period 1970-1978 were calculated using the balancing equation approach (see Table 7.2).[9] In addition, estimates for 1971-1981, presented by K. Srinivasan to the Demography Association of India Conference in 1983, are included.[10] These were derived using vital rates for the year 1976 and adjusting them using population totals of the 1981 census.

The two sets of estimates agree in sign for all states except T.N., and differ significantly in magnitude only in the case of M.P. and Orissa. However, despite these discrepancies, it seems that during the 1970s, Bihar and Karnataka became the foremost gaining states, followed by Orissa and M.P.. Among the losing states, Punjab and Haryana seem to have insignificant outflows of population, and Kerala is the only state that experienced significant outmigration. The reasons for some reversals over the previous decade are stated in the concluding section.

A long term view of net migration patterns indicates that Karnataka is the only state that has attracted population since 1951, whereas Kerala is the only state that has consistently lost population during the last three decades. In addition, the outflow of population from the Punjab has been decreasing over time.

Determinants of Interstate Migration

Did the structural changes experienced by the economies of high growth states such as Punjab and Haryana (discussed in chapter two) alter the demand for labor and if so, was it perceived by migrants and did it stimulate migratory flows from areas of low labor demand to regions of high demand? Did those Indian states where growth was most extensive

TABLE 7.2

Average yearly net migration rates (per 1000), 1970-1981

State	1970-1978	1971-1981
A.P.	1.33	0.87
Bihar	5.03*	3.02
Haryana	-0.04	na
Karnataka	4.71	3.97
Kerala	-1.43	-1.92
M.P.	1.52	0.31
Orissa	1.66	0.53
Punjab	-0.37	-0.65
T.N.	0.93	-0.41
U.P.	0.62	0.86

*= 1970-1976 only
na= not available
Source: (first column) derived from various tables in Government of India, Vital
Statistics Division, *Sample Registration of Births and Deaths in India*, yearly volumes
1964-1978, New Delhi
(second column) K. Srinivasan, "India's Demographic Trends", unpublished paper
presented at the Demography Association of India conference, February 1983.

act as a pull on the unemployed population from the remaining parts of
India?

Table 7.3 contains a ranking of states by the total growth achieved
during each decade with their corresponding migration rates. There is
clearly no relationship between the two variables. During the 1960s, both
the highest and lowest growth states experienced relatively large
outmigration, whereas the remaining states often alternated in the direction
of migratory flows. Even though the 1970s are characterized by less
fluctuation, a pattern is hardly recognizable. In the first decade, the
correlation coefficient of growth regressed upon migration is .004,
whereas in the second decade it is -.36. In both time periods then, there is
evidence of no relationship between economic growth and net interstate
population flows. This is illustrated most clearly by the experience of the
high growth states: Punjab lost population through migration whereas
Haryana experienced inmigration during the 1960s and outmigration in the
1970s. Yet, in both decades, states with mediocre rates of growth attracted
population.

Since it is misleading to observe growth in isolation of the levels of
income per capita, these data are presented in Table 7.4. It is possible that
economic change (or growth) is not perceived by the migrant, and that the

level of economic activity (or per capita income) is more relevant in the decision to migrate. However, it is again clear from simple observation that there exists no pattern between the levels of income per capita and migration rates in the agricultural states. Regression analysis confirms this: migration regressed upon income in the 1960s yields .013 and .221 in the 1970s.

This lack of relationship between migration and economic variables may have resulted from the comparison of data from the same decade. This cannot account for any lag in the migration process. Such a lag may exist because time elapses between the first perception of economic opportunity and the migrant's response to it. Although Eckhaus (1984) points out that it is not uncommon to view ten years as a short lag in questions of labor supply, in the case of agricultural work (which constitutes the bulk of short distance employment "pull"), lags are usually seasonal or yearly in duration and thus did not warrant combining of growth of the 1960s with migration in the 1970s.[11] Given that migration data is usually presented by decade, it was impossible to incorporate lags of one year, and inappropriate to assume them to be ten years in duration.

Notwithstanding the issue of lags, the striking conclusion that economic growth and income in the agricultural states of India are not determinants of interstate migration indicates the desirability of broadening the scope of variables which might have an influence on migration patterns. Given the particular conditions of predominantly agricultural economies, three possible influences are discussed: the prevailing agricultural technology with its labor absorbing or displacing properties, the population density and the perceived opportunities in the industrial regions.

The Source of Growth

It was stated in chapter two that economic growth in the predominantly agricultural regions stems from either the increased use of already existing inputs, or the introduction of new, more efficient ways of production. In the latter case, the labor requirements usually decrease as greater efficiency permeates production methods. However, as pointed

TABLE 7.3

Economic growth and migration

state	(1960s) growth	migration	state	(1970s) growth	migration
Punjab	3.1	-1.87	Punjab	2.6	-0.37
Haryana	3.0	0.74	Haryana	1.9	-0.04
Karnataka	2.5	0.31	Bihar	1.1	5.03
Orissa	2.0	0.32	A.P.	1.0	1.33
Kerala	1.4	-1.55	Orissa	1.0	1.66
A.P.	1.2	-0.32	U.P.	0.6	0.62
U.P.	0.7	-1.28	T.N.	0.6	0.93
T.N.	0.4	-0.30	Kerala	0.5	-1.40
M.P.	0.04	0.83	M.P.	0.2	1.50
Bihar	-0.9*	-1.01	Karnataka	-0.7	4.71
$r^2 = .004$			$r^2 = -.362$		

* refers to growth over 1960-61 through 1969-70
Source: Tables 2.2, 7.1 and 7.2.

TABLE 7.4

Levels of income per capita and migration

state	(1960s) income	migration	state	(1970s) income	migration
Punjab	409	-1.87	Punjab	1225	-0.37
Haryana	384	0.74	Haryana	969	-0.04
T.N.	343	-0.30	Karnataka	661	4.71
A.P.	293	-0.32	Kerala	637	-1.40
Kerala	279	-1.55	A.P.	618	1.33
Karnataka	272	0.31	T.N.	598	0.93
U.P.	261	-1.28	Orissa	504	1.66
M.P.	260	0.83	U.P.	503	0.62
Orissa	241	0.32	M.P.	489	1.50
Bihar	211	-1.01	Bihar	425	5.03
$r^2 = .013$			$r^2 = .22$		

note: income refers to the average total income per capita, in constant Rs., for 1960-61
to 1970-71, and 1970-71 to 1980-81
Source: income statistics derived from Table 2.1, migration statistics reprinted from
Table 7.1 and 7.2.

out in chapter three, there are exceptions to this since the particular technology associated with the green revolution may be either labor absorbing (as in Punjab) or labor displacing (as in Haryana). In the case of growth stemming from an increase in inputs, the labor demands increase until the point of decreasing marginal returns is reached. Hence, for the rational and informed migrant, the perceived economic opportunity might depend upon the source of growth rather than the growth itself. However, it is not always the case that growth stemming from a mere increase in inputs constitutes a stronger pull on the migrant than growth stemming from technological change. The net result of offsetting tendencies must be considered: technological change is sometimes labor intensive (as in the first decade of the green revolution in Punjab), and it may stimulate production so that diffusion of growth into other sector occurs and total labor use increases. Although perhaps slightly unrealistic, it is assumed that the rational migrant's decision to relocate incorporates all these considerations.

Does economic growth *derived from increases in inputs* stimulate migration? According to the data presented in Table 2.4, the states that experienced an increase in land use were Kerala and M.P. Of these, Kerala was a net loser of population, whereas M.P. absorbed migrants. With respect to growth from an increased use of labor, Punjab and Kerala absorbed the greatest number of workers in agriculture, yet they were also the states with the largest degree of outmigration. It is clear that migrants were not attracted to regions where input increases contributed significantly to economic growth.

With respect to "intensive" sources of economic growth (i.e., innovative technology), it should be recalled that Punjab adopted most technology in the form of tractors, irrigation and fertilizers. Although it may be inferred from neoclassical theories of mechanization that the labor component tends to decrease when capital inputs increase, it was clear from chapter three that the prevalence of biochemical aspects of the green revolution actually resulted in an increase in agricultural workers. This increase in workers was concomitant with net outmigration. Haryana's agriculture was also characterized by extensive technological innovation, which resulted in a displacement of primary sector workers while migrants flowed into the state. This evidence indicates that the source of growth in agriculture, the largest sector in the states under study, is not a determinant of state level migration patterns in India.

Population Density

It is plausible to argue that the greater the number of people residing in a given region, the greater the competition for employment, housing, facilities, food, etc. Hence, under conditions of insufficient supply of these goods and services, it is logical to assume that population density acts as a push, causing people to relocate to regions where the competition is smaller. Among the ten agricultural states of India, this negative relationship between population density and migration holds in two states only: Kerala is a large loser of population and has by far the greatest population density (435 people per square kilometer in 1961) and M.P. is a population absorbing state where the population density in 1961 was a mere 73 per sq km.[12] However, this negative relationship is not so clear for the remaining states. In fact, the regression of migration rates on population density in 1961 and 1971 yield a correlation coefficient of -.55 and -.17 respectively, both sufficiently low to indicate the lack of relationship.

Growth in the Industrial States of India

The ten agricultural states do not constitute a closed system since migrants also relocate out of the country or into the predominantly industrial states. Given the insignificance of international migration, the relatively industrial states of Gujarat, Maharastra, Rajasthan and West Bengal are the principal possible destinations of population flows outside of the ten agricultural states and on that basis warrant inclusion in the study. Although it was suggested above that economic growth does not correlate with interstate migratory patterns *among the agricultural states*, a different relationship between migration and economic variables may emerge when the absorbing economy is predominantly industrial. Because the nature of work in industrial regions tends to be largely non-agricultural, the perceived economic opportunity in Maharastra, Gujarat, Rajasthan and West Bengal might act as a pull on migrants from the agricultural states.

Some income comparisons are in order. The average percent change in income per capita experienced during the 1960s in the industrial states was 12.6, compared with 20.0 in the agricultural states, and 6.1 and 9.4, respectively, in the 1970s.[13] The average in the agricultural group is biased upward as a result of the experience of Punjab and Haryana, whereas the relatively low growth in Rajasthan biases downward the industrial states' average. The average income per capita in the agricultural states has been lower than in the industrial states during the 1960s and 1970s, culminating in the following difference in 1971 and 1981: in Rs., 624.3 and 738.5 (in 1971) and 694.9 and 789.3 (in 1981), respectively. If economic opportunity is in fact embodied in high levels of income per capita, and is viewed as such by the migrant, it follows that the industrial states provide a greater pull on the migrant. According to migration statistics for the industrial states, Gujarat, Maharastra and West Bengal were net population gainers (average annual net migration rates during 1960s were .33, 1.84 and .87 respectively), whereas Rajasthan was a loser (-.95).[14] It therefore seems that these industrial states experienced mostly inmigration even though the agricultural states as a whole were characterized by greater economic growth (migration regressed on economic growth yields .53). However, the industrial states surpassed the agricultural states in their income per capita and the correlation between migration and income per capita in the former group of states is .95. This suggests that migrants tend to view the level of income in a predominantly industrial economy as providing sufficient opportunity to warrant interstate movements.

Further support for this is provided by evidence pertaining to the origin of the population that migrated into the industrialized states, as well as destination of population flows from agricultural states. According to Indian government estimates, the population absorbing industrialized states received the largest proportion of migrants from agricultural states: Gujarat, Maharastra and West Bengal absorbed 32, 65 and 42 percent, respectively, of their male migrants from the ten agricultural states under study. With respect to the spatial destination of migrants from the agricultural states, the following distribution is noted: from Punjab, 39 percent moved to the industrial states and the Union Territory of New Delhi, compared to 25 from Kerala and 53 from U.P. (Government of India 1979: 84-89). A breakdown by state reveals that proximity is an important consideration in migration, so that the closest industrial states tend to absorb the largest number of migrants.

Migration and the Labor Force by Sector

Are migrants reflected in the changing labor component of the three sectors that undergo change during long term economic growth? By observing transformations in the sectoral distribution of the labor force in the outmigrating and inmigrating states, inferences can be made about which sectors absorb or lose the migrating population.

A comparison of average migration rates with the percent change in the sectoral distribution of the labor force is presented in Table 7.5. States were ranked in descending order by their net migration rates, then grouped into three categories: the outmigration, insignificant migration and inmigration states.

TABLE 7.5

Migration (by category) and changes in the labor force by sector, 1961-1971

ave. migration	% change in labor force by sector		
	I	II	III
(out.) -1.43	14.86	-20.41	-12.31
(ins.) 0.01	6.31	-11.19	-0.55
(in.) 0.79	-3.23	-3.28	14.98

note: out.= outmigration states, ins.= insignificant migration states, in.= inmigration states
Source: derived from Tables 7.1 and labor force tables in Registrar General, Government of India, *Population Census 1961, 1971*, New Delhi.

The outmigration states lost labor from services and industry and gained labor in agriculture. Included in this group are Punjab, Kerala, U.P. and Bihar. Two of these are states where high rates of agricultural growth occurred, indicating that where growth in agriculture was experienced, the population exodus was *not* from this high growth sector. This coincides with the evidence of labor absorption in Punjabi agriculture presented in chapter three.

It might be argued that the primary labor force of the outmigration states did not decrease during the 1960s since the migrant tends to be an individual outside the labor force and in search of employment. An observation of the change in rural residency over the same time period shows that all states experienced a decrease in the rural population as a percent of the total.[15] More importantly, the high growth states had the *least* decrease in rural residency. It thus seems that out of state migration did not necessarily originate in the rural areas. Since there was a simultaneous decrease in the labor force in manufacturing and services (without a significant increase in the income from these sectors which might indicate labor saving innovation), it is suggested that where agricultural growth occurred, out of state migration was largely from non-rural areas and the non-agricultural sectors of the labor force.

Meanwhile, population absorbing states experienced a decrease in labor from primary and secondary sectors, and an increase in service employment of 14.98 percent. The fact that these same states have had the lowest growth in agriculture (with the exception of Haryana), coupled with a decrease in the agricultural labor force, suggests that the inmigrating population might have largely been absorbed by services. This is the only group of states that exhibits the typical Kuznetsian pattern of migration and structural transformation of the economy: inmigration to satisfy the new demand for labor in the tertiary sector, coupled with a decrease in the demand for workers in agriculture, as well as a slight decrease in industrial workers. In fact, census data pertaining to the industrial structure of working migrants reveals that in Haryana 42.2 percent of migrants worked in the tertiary sector, compared to only 17.4 percent in M.P. Even if Orissa were included in this category (and according to the National Research Council (1984) estimates, it is a population absorbing state), migration absorption in services is relatively high: 26.8 percent. In view of the fact that many migrants are not employed at the time of census enumeration (in the agricultural states, ranging from a high 43.6 percent in A.P. to a low 21.0 in U.P.), these numbers are relatively high (Government of India 1979: 50-51).

Conclusions

The evidence seems to indicate the lack of a relationship between net interstate migration and several variables that might affect the push or pull of population, such as growth or income levels in the agricultural states, the source of growth or population density. This may be the result of the inappropriateness of aggregation at the state level in the Indian setting, perhaps due to the large interstate differences in income levels which were not present in the states studied by Lee, Kuznets and Eldridge (1964). In the Indian case, a regional or district level study might indicate a positive relationship between some of the above variables since it would isolate the high growth regions within a state, such as Ludhiana in the Punjab.[16]

However, the evidence presented above does indicate a positive relationship between predominantly industrial economic performance and inmigration, suggesting that interstate migration will more likely occur when the final destination is associated with non-agricultural employment opportunities. This is similar to the conclusions drawn from the analysis of the relationship between the interstate migratory pattern and sectoral changes in the labor force, namely:

(1) migration does not happen in response to economic opportunity in the high growth sector *if* that sector is agriculture, regardless of the source of growth,

(2) the perceived labor force demands of the secondary and tertiary sectors are relevant in the decision to migrate, even if these sectors are not growing as rapidly as agriculture.

This suggests that migrants will relocate for urban wage employment because it exerts a sufficient pull for interstate migration. Those states that attracted migrants were not the states in which agricultural labor was increasing, but instead were those in which the manufacturing and service sectors were perceived as capable of absorbing labor, even though they were not necessarily the highest growth sectors. Interstate movements entail commitment of greater magnitude and permanence and would not be undertaken for less than the possibility of wage employment in a "modern" sector. It is seasonal migration, usually reflected on an inter- and intra- district level, that occurs in response to agricultural demand for labor.

These conclusions are relevant in the explanation of the apparent turnaround in migratory flows to and from Bihar. During the 1960s, this state was losing population, whereas during the 1970s, together with Karnataka, it became a major population absorber. Indeed, there is evidence that during the late sixties and early seventies, an industrial expansion occurred as the mineral, engineering, chemical and other industries proliferated (reflected in the increase in income from manufacturing from 10.6 percent to 21.3 percent during 1969-70 to 1978-79).[17] With respect to the industrial occupation of migrants, some 31 percent of the males were absorbed in the secondary and tertiary sectors, compared to 22 percent in agricultural activities (Government of India 1979: 50-51). Thus, non-agricultural employment might have attracted migrants during the course of Bihar's industrialization process.

Notes

1. This is a three volume study entitled *Population Redistribution and Economic Growth, United States:1870-1950*. The first volume (by Lee et al.) was published in 1959, the second (by Kuznets et al.) in 1960 and the third (by Eldridge et al.) in 1964. A comprehensive review of these three volumes is presented in L.O. Stone (1968), "Population

Redistribution and Economic Growth, United States: 1870-1950, a Review Article" in *Demography* 5.

2. Recent studies of the U.S. economy and migration include Michael Greenwood (1981), *Migration and Economic Growth in the United States*, New York: Academic Press, and relevant articles in Alan Brown and Egon Neuberger (eds.), (1977), *Internal Migration: A Comparative Perspective*, New York: Academic Press.

3. As a result, earlier theories of development viewed internal migratory movements favorably. The withdrawal of cheap manpower from agriculture and its availability for industries was deemed socially beneficial since resources were shifted from where their marginal product was low to where it was higher. See W.A. Lewis (1954), "Economic Development with Unlimited Supplies of Labor" in *Manchester School of Economic and Social Studies* 20, and J.C. Fei and G. Ranis (1961), "A Theory of Economic Development" in *American Economic Review* 51.

4. Jolly (1970) was among the first to point out what is considered obvious in the current literature, namely that internal migration is not entirely socially beneficial since it exacerbates structural imbalances between the rural and urban regions ("Rural-Urban Migration: Dimensions, Causes, Issues and Policies" paper presented to the Conference on Prospects for Employment Opportunities in the 1970s, Cambridge University).

5. Theories pertaining to selectivity of migrants incorporate such factors such as age, sex, educational and occupational differentials. Shaw contains a review of the literature on selectivity (1975, *Migration Theory and Fact*, Philadelphia: Regional Science Research Institute).

6. Kingsley Davis (1951), *The Population of India and Pakistan*, Princeton: Princeton University Press.

7. The data in this paragraph are taken from Gosal and Krishan (1975), pp. 195-198.

8. These data, as well as those in the following paragraph, are derived from Premi (1982), Table A-6.

9. The balancing equation method of indirect intercensal migration estimation consists of treating migration as a residual after subtracting the natural increase in the population from the total increase at two points in time. The natural increase was estimated using vital statistics from the following sources: Government of India, Office of Registrar General and Census Commissioner, Ministry of Home Affairs, *Sample Registration Bulletin* (yearly volumes 1965-1975), and *Vital Statistics in India*, New Delhi, as well as Vital Statistics Division, *Sample Registration of Births and Deaths in India* (yearly volumes 1964-1978), New Delhi.

10. The author is indebted to Mari Bhat of the University of Pennsylvania for the use of this unpublished material.

11. R.S. Eckhaus (1984), "Some Temporal Aspects of Development" in *World Bank Staff Working Paper* 626, Washington.

12. The data pertaining to population density in 1961 are taken from Government of India, Central Statistical Organization, *Statistical Abstract of India 1963-64*, New Delhi. Population densities in 1981 are presented in Table 1.2.

13. For the source of the data pertaining to industrial countries, see Table 2.1.

14. Migration statistics for industrial states are also derived from Premi (1982), pp.140-141.

15. From 1961 through 1981, there has been a clear increase in the urban population relative to the total in all states. See various *State Statistical Abstracts*.

16. An excellent district level migration study of Ludhiana district can be found in Oberai and Singh (1980).

17. Economic data on Bihar are taken from: Government of India, Ministry of Information and Broadcasting (1983), *India*, New Delhi, pp. 478-479.

8

The Mode of Production in Indian Agriculture

Recent models of agricultural development can be classified into two groups according to the traditions from which they emanate: those with roots in neoclassical theory, associated with scholars such as Johnston, Mellor, Ruttan, Schuh, or Schultz, and those with origins in the tradition of radical political economics, espoused by writers such as Amin, Beckford, deJanvry, Emmanuel and Frank. Within the radical group of theories there exists a major subdivision with respect to agricultural issues: pure Marxist analysis and dependency theories. Scholars in the Marxist tradition have addressed the role of agriculture in development primarily as it relates to the western and Soviet experience. Lenin, Kautsky, Bukharin, Preobrazhensky and others debated the relative importance and role of agriculture during the industrialization process. The neglect of the Third World ended with the publication of Baran's work (1957), which underlies much of modern neomarxist analysis as well as dependency theories. The dependency theorists primarily focused on Latin America, whereas the nature of Asian agriculture was the theme of scholars such as Avineri (1969) and Melotti (1977).

The major contributions made by radical agricultural studies to development theory and practice in the past two decades may be summarized as follows. First, the individual farmer was not viewed as the central cause of poverty but rather as an insignificant component in the problem of agricultural development. Schultz's theory (discussed in chapter two), pertaining to the rationality of the farmer who merely needs correct inputs to increase output and thereby decrease rural poverty, was criticized on grounds that the root of agricultural stagnation lay elsewhere.

153

Specifically, the focus shifted to factors exterior to the farmer, among which the most important was the nature of linkages between the village level producer and the global economy. Second, radical political economics rests on the belief that a social order characterized by an unequal distribution of power and assets will influence institutional arrangements which in turn influence the production capacity of the individual farmer. The role of power structure is viewed as crucial not only within the village, but also within the nation as well as among countries. Third, Marxist literature stressed the uniqueness of regions with different historical processes and social structures, and thereby rejected some models of agricultural development which prescribed the same medicine with no regard to particular local conditions.

From such radical economic writings a Marxist literature on Indian agriculture emerged which reflected frustration at the inability of mainstream development strategies to deal effectively with worsening rural poverty. In addition, the unequal distribution of benefits of the green revolution among the rural population and the increased polarization of the rural classes resulted in extensive Marxist analysis of Indian agricultural development. The most important research took place within the context of what became known in Marxist literature as the "Indian debate." During the last twenty years, a theoretical and empirical debate emerged with regard to questions of political economy, mostly in the journals *Economic and Political Weekly*, *Frontier*, and *Social Scientist*.[1] The debate, which consisted of vehement argumentation among Indian and foreign scholars, is Marxist in theory, methods and historical analysis, although there is ample disagreement about what constitutes a "true" Marxist approach in each of these areas. Although it is difficult to combine all these exchanges into a single debate, it is clear that the major issues focused on the identification of the mode of production in India (was it feudal, semi-feudal or capitalist?), how this mode was manifested, how it can be measured, what are its ramifications, etc. International Marxist literature recognizes the importance of these exchanges and places them on par with the Latin American Marxist contribution insofar as they both address a broad range of issues in the effort to understand the development process.[2]

The nature of the analysis in this study thus far has been along the lines of the western neoclassical tradition. Given that neither the neoclassical nor the radical approach is invariably right or wrong but both have merits and demerits, the study of Indian agriculture would be incomplete without the introduction of another perspective. In an effort to fill this gap, this chapter contains a study of the agricultural regions of India studied from a Marxist perspective using the concept of the mode of production. An attempt is made to empirically ascertain the nature of the mode that prevailed in the ten agricultural states during the 1960s and 1970s. In addition, the relationship between the mode of production and economic growth is identified, as well as the link between the mode of production and welfare of the society. These last two sections represent an effort to integrate aspects of Marxian analysis and Kuznets's theory of modern economic growth, especially with respect to their similar views on the relationship between technology, economic growth and welfare.

The Capitalist Mode of Production: Theory

The concept of a mode of production is a useful tool in trying to characterize an economy under investigation. This tool can provide a way of classifying a society which thereby serves the same general function that all models do, namely, to clarify and highlight the most salient elements of a society. In the Marxist model, the mode of production is composed of two distinct entities: the relations of production and the forces of production.[3] The former describes the relationship between the owners of the means of production and the producers, who, through some form of agreement, work for the owners. The relations, then, express both the structure of ownership in an economy, as well as the relationships between social groups which emerge as a result of that structure. On the other hand, the forces of production relate to the state of technology, as well as the stock of capital (buildings, tools and machinery) and labor force available to the economy.

Among Marxists there is a general consensus concerning the characteristics of a *capitalist* mode of production. These defining characteristics include, with regard to the relations of production, the *existence of free labor*, the *separation of owners from the means of production* and the existence of *private capital accumulation*. With regard to the forces of production, methods of production must be sufficiently advanced to allow for production of *commodities in excess of consumptive needs* (generalized commodity production) and thus allow for the marketing of output. In this context, workers who own no income-generating property except their labor power (and are thus free), sell that labor for wages to the owners. These owners of the means of production then mobilize the forces of production in an effort to earn revenue in excess of their consumptive needs.

While many authors agree on this outline of an analysis of the mode of production, much disagreement has arisen in the course of identifying the mode which best characterizes agricultural production in India. Two questions seem to be at the center of this controversy: first, how much importance should be assigned to each of the attributes that identifies a mode and secondly, how can these attributes be empirically identified and measured. As part of our effort to characterize the agricultural mode of production, we will refer to the issues and participants in this debate in order to clearly identify the position we have adopted in relation to other students of the problem. Furthermore, it should be noted that the choice of the four characteristics of capitalism (free labor, alienation from the means of production, production for the market and private capital accumulation), emanates not only from formal Marxist theory, but also from exchanges which are part of this Indian debate.

Free Labor

In a capitalist mode, labor power is a commodity which is sold for wages in the labor market. Capitalism is thus different from pre- or non-capitalist modes insofar as class relations are not characterized by personal dependence (as in the ancient and feudal modes),[4] but instead are founded on exchange. "Formally, the relationship [between the workers and owners] is the free and equal one of those who exchange" (Marx in Howard and King, 1976: 87). Money wages, which are determined in the labor market, are payment for the labor power which workers must be free to dispose of, at will, through the labor market. For the optimal efficiency of this market, there must be no constraints on the demand and supply of labor. Such constraints might entail extra-economic impediments to industrial as well as regional mobility, both of which act to limit the alternatives open to the labor force. In a capitalist mode, labor must be free to relocate in response to economic or other incentives since without these alternatives, labor is effectively unfree.

The inclusion of regional and industrial mobility into the concept of free labor in India represents an elaboration of a view presented by Patnaik (1971). She was the first among the participants of the debate on the Indian mode of production to introduce the issue of the "unfreedom" of Indian agricultural workers resulting from the lack of other employment opportunities: "In the absence of alternative job opportunities, the [agricultural workers] are effectively tied to agriculture as the main source of livelihood" (Patnaik 1971: 124). In other words, their industrial mobility is restrained by the economic condition of India in the 1950s and 1960s which did not offer alternative economic opportunities. Her view was considered unorthodox by those who strictly applied orthodox Marxist definitions and concepts to the Indian setting. Chattophaya (1972), for example, interprets labor freedom as entailing only freedom from land ownership, and on those grounds identifies Indian labor as non-free, therefore non-capitalist. Labor's physical or industrial mobility was not included, and therefore his analysis fails to encompass the wider aspects of labor freedom. Unfortunately, Patnaik failed to respond and the issue of labor freedom does not reappear in the Marxist literature on Indian agriculture until 1981, when two of the participants of the debate reintroduced the issue. Rudra (1981) argued that West Bengal is characterized by labor immobility and takes this as an indication of the absence of capitalism. Omvedt (1981) presented data from Maharastra indicating seasonal labor mobility in response to agricultural employment opportunities, thus identifying a capitalist characteristic. Both of these are, however, secondary results of isolated studies conducted on a limited scale and provide no all-India conclusions nor a basis upon which to draw them. In sum, the debate on the Indian mode of production remained without a consensus on whether agricultural labor is free in India.

Separation of Workers From the Means of Production

Another dimension of the freedom of the labor force entails the disassociation of the worker from the ownership of a means of production. According to Marx:

> ...On the one side the possessor of value or money, on the other, the possessor of the value-creating substance, on the one side, the possessor of the means of production and subsistence, on the other the possessor of nothing but labor power, must confront each other as buyer and seller. The separation of labor from its product...(was) the starting point of capitalist production.
>
> (Marx 1970: 570)

In Marx's theory, this separation of the worker from the means of production is salient to the capitalist mode insofar as it determines the ownership and thereby the income structure of a society. Workers, who own no income-generating property, receive wages for their labor. They cannot reap any "surplus value" but can only be the source of that surplus extraction. Given that this minority class of owners of the means of production are motivated to increase their surplus, there exists a divergence of interests and income between them and the workers.

On this issue, the Indian debate reflects a tendency to adapt Marxist definitions to the particular conditions of the Indian setting. In the late sixties and early seventies, free agricultural workers were defined as only those that owned no land whatsoever, and thereby satisfied Marx's requirement that "the worker must be separated from the land" (Hobsbawm 1975: 67). On these grounds, as was mentioned above, Chattophaya (1972) refuted the existence of a capitalist mode of Indian agriculture, since landless labor is by no means the predominant agrarian group. Omvedt, in 1981, argued that in the context of the Indian agrarian structure, it is meaningless to speak of a holding smaller than 2.5 acres as property capable of generating even subsistence income. Therefore, she accommodated the Marxist definition and classified those workers who are either landless (or own unviable holdings and must therefore hire out their labor for wages) as owning no income-generating property. Omvedt's contribution represents a significant step in the direction of reevaluation of issues that the early stage (1962-1972) of the debate addressed, but with the added benefit of the data base that was not available to the earlier participants. Among these early studies, both Rudra (1970) and Gupta (1962) used the existence of wage labor as one of the characteristics of the capitalist mode, and they reached opposing conclusions. Gupta, basing his study on 1953-54 data, identified capitalism on the basis of a proliferation of farms which used hired labor for wages, the existence of which indicated a population separated from income-generating property. Rudra, on the other hand, claimed there was no evidence of the capitalist mode in Punjab in 1968 because the characteristic features of capitalism, among them the existence of free wage labor, were not statistically correlated with each other.[5] Although both of these studies have been criticized on grounds of questionable methods,[6] they at least recognized the importance of wage labor with no income-generating property in the determination of the capitalist mode.

Capital Accumulation and Production for the Market

The interaction of the relations of production with the forces of production is evident in the process of private capital accumulation which is itself dependent upon production for the market. According to Marxist theory, capitalism entails a system of "extended commodity production" in which production by the hired worker exceeds the consumption of the employer. The production of commodities of greater value than the cost of the production process yields a "surplus value," which is then realized by the sale of commodities for money. This money is in turn reinvested into new means of production since "capital strives after the universal development of productive forces, and then becomes the prerequisite for a new means of production" (Marx in Howard and King 1976: 111).

These two characteristics of capitalism, capital accumulation and production for the market, do not exist in economies at different historical stages. According to Marxist theory, in a pre-capitalist economy, simple commodity production takes place, in which "independent artisans and farmers" own their means of production and market products themselves. Income accrues in direct proportion to the labor input, only income from work exists, there is no profit and there is then no capital accumulation. This comparison with pre-capitalist economies serves to show the crucial importance of capital accumulation for the capitalist mode, as well as production in excess of consumptive needs, since without the capacity for generalized commodity production, the owners could not appropriate "surplus value" nor receive income from their ownership.

The Indian debate reaches a tenuous consensus on the importance of capital accumulation. Two groups of scholars approached the issue from different angles, largely ignoring each other's work. One group addressed the accumulation of capital that occurred during colonial times through the siphoning of surplus to the metropolitan colonial centers, thus placing India in the international capitalist sphere and including it into the parent debate between Laclau and Frank.[7] The second group of scholars limited themselves to the national economy. Of these, Patnaik was the most vociferous defendant of the position that capital accumulation and capital reinvestment in agriculture were the principle distinguishing features of the capitalist mode. Her argument rests on the fact that surplus appropriation and hence capital accumulation existed during colonial times also, but that in post-Independence India it distinguished itself insofar as farmers reinvested into their land through the accumulation of machinery and other technological inputs.[8] This investment was stimulated by the expansion of post-Independence markets which acted to increase the profitability of farming. Although she was criticized as describing little other than "a higher degree of technological development" (Chattophaya 1972) and "merely looking for industrialization of agriculture" (McEacheron 1976), she was applauded for identifying the importance of capital accumulation. Sau (1973), Rao (1970) and Rudra (1970) independently came to similar conclusions on the importance of capital accumulation,[9] although more so on the grounds of consistency with the theoretical Marxist model than for its contribution to capitalist production in Indian agriculture.

Indeed, since the Marxist description of capitalism is so intricate and encompasses many variables, the attribution of unique importance to a single feature is a grave oversimplification. Both McEacheron and Chattophaya are guilty of this by virtue of their objection to Patnaik's straying from the narrowest definition of the relations of production, as is she for making capital accumulation the necessary and sufficient condition of capitalist existence. It is furthermore unclear how capital accumulation can be separated from generalized commodity production. Yet, although there has been general consensus among the participants that production for the market exists in India (introduced during colonial times as a result of the stimulation of trade), its theoretical link to capital accumulation is not made explicit. Chattophaya and McEacheron, among others, give production for the market foremost importance in the determination of the capitalist mode. For Rudra, on the other hand, it was only one of six characteristics, although the only one that was highly correlated with others. This position of interdependence among the variables that describe the capitalist mode is adopted in this study.

The Capitalist Mode of Production: Empirical Findings

Free labor, the separation of workers from the means of production, capital accumulation and production for the market are thus the defining characteristics of the capitalist mode. These have been operationalized through the identification of indicators that can be measured in rural India. Although the existence of a capitalist mode may be inferred in the states in which these indicators are present, it must be pointed out that several modes of production may coexist in a region at the same time, since demarcations between modes are not discrete and absolute.[10] In a subcontinent as diverse and large as the Indian, with differing impacts of differing historical conditions, including diverse colonial experiences, one expects not only different modes to coexist in proximate regions, but even variations of the same mode. In this study, the concern is with the identification of *tendencies* within the states, rather than the search for absolutes which, given the nature of social systems, cannot exist.

Free Labor

For the purposes of this study, labor freedom is defined as regional and industrial mobility, and the indicator of its existence is the total interstate migration experienced by the states. The degree of interstate migration, measured by the sum of intercensal, current and lifetime migration (see chapter seven) provides an indication of the general movement of the population, and thus the facility with which workers shift location in response to opportunities in another region.

It is evident from Table 8.1 that Punjab and Haryana distinguish themselves from the other states with respect to population mobility. In these states, 27.8 and 38.1 percent of the 1971 populations are interstate migrants, as compared to the remaining states in which migrants range from 8.1 to 14.9 percent of the population. Thus, it seems that in Punjab and Haryana, labor is free to relocate in search for employment opportunities to a greater degree than elsewhere, and by that criteria these states are relatively more capitalist.

TABLE 8.1

Total migrants as a percent of total population, 1971

A.P.	8.1
Bihar	11.2
Haryana	38.1
Karnataka	10.2
Kerala	13.2
M.P.	10.7
Orissa	14.9
Punjab	27.8
T.N.	9.2
U.P.	8.9

Source: calculated from Premi (1982: 140, Table A-6).

Separation of Workers From the Means of Production

In rural areas, land ownership might be considered the principal form of income-generating means of production, so landlessness may be considered an indication of the extent to which workers are separated from the means of production. Although in India, some holdings are so miniscule that they scarcely satisfy subsistence needs, the owners of these marginal plots have not been included in the measure of landless agricultural workers appearing in Table 8.2 as a result of data constraints.[11] Thus the data underestimate the proportion of workers

TABLE 8.2

Landless labor households as a percent of total agricultural labor
households, 1974-75

A.P.	60.9
Bihar	43.4
Haryana	83.2
Karnataka	53.4
Kerala	13.3
M.P.	47.3
Orissa	37.2
Punjab	56.2
T.N.	63.8
U.P.	43.0

Source: calculated from Omvedt (1981), Table 10. Her data in turn are calculated from
the Rural Labour Enquiry (1978), Table 2.3(b).

that are effectively separated from the means of production. In Punjab and
Haryana, 91.4 and 83.2 percent of the agricultural worker households
have no source of income other than their labor. These values must be
compared to an average 45.3 percent in the remaining eight states. This
high degree of worker alienation may be taken as an indication of the
existence of capitalist relations of production in Punjab and Haryana.

Production For the Market

Farm size has been adopted as the indicator of the existence of
generalized commodity production. As farm size increases, production
capacity exceeds subsistence needs, resulting in the marketing of the
excess. With increased profitability of agricultural production due to the
green revolution, the right conditions were created to stimulate potential
market oriented farmers into capitalist production. Therefore, on the basis
of production in excess of consumptive needs, we can use farm size as a
proxy for production for the market. Although the green revolution made
capitalist farming more likely for all farm sizes, cash crop production was
less likely for the small farmer.

TABLE 8.3

Percent of area under middle size (10-20 acres) and large size (20 acres and above) operational holdings, 1971

A.P.	53.2
Bihar	39.7
Haryana	64.1
Karnataka	34.2
Kerala	18.6
M.P.	68.5
Orissa	31.6
Punjab	56.2
T.N.	30.2
U.P.	2.6

Source: various tables in Naidu, I.J. (1975), *All India Report on the Agricultural Census 1970-71*, New Delhi: Ministry of Agriculture and Irrigation, Government of India.

Land concentration[12] patterns in ten states were studied by analyzing the proportion of land under cultivation by large and middle farmers, i.e., by holders of 10 to 20 acres, and 20 acres and above.[13] 1971 was chosen as the year of observation since it represents a stage of the green revolution at which time the most enterprising farmers have already been absorbed into generalized commodity production. The empirical findings presented in Table 8.3 show that land concentration was highest in M.P. and Haryana (68.5 and 64.1 percent, respectively), followed by Punjab (56.2 percent). On the other extreme, states such as U.P. and Kerala have very little area under middle and large size holdings (2.6 and 18.6 percent, respectively). Therefore, generalized commodity production is more prevalent in M.P., Haryana and Punjab than in the other states.

Capital Accumulation

Mechanization is an obvious indicator of capital accumulation since adoption of mechanical inputs in agricultural production shows the investment of capital for the purpose of increased production and profits. Table 8.4 contains data on the change in tractor ownership, aimed to serve as an indication of the degree of mechanization during the initial years of the green revolution (1966 through 1972).

TABLE 8.4

Percent change in tractor ownership, 1966-1972

A.P.	116.4
Bihar	162.6
Haryana	279.4
Karnataka	148.8
Kerala	258.9
M.P.	99.0
Orissa	169.9
Punjab	298.3
T.N.	64.7
U.P.	172.2

Source: reproduced from Table 2.9.

It is evident that Punjab and Haryana surpassed the remaining states with respect to the level of mechanization (298.3 and 279.4 percent increase in tractor ownership), indicating a greater degree of capital accumulation than occurred elsewhere.

Conclusions From Empirical Evidence

It is clear from the above section that the features of the capitalist mode of production are present in Punjab and Haryana to a greater degree than elsewhere. These states are characterized by an environment in which medium size plots prevail, on which farmers cultivate using wage labor and accumulate capital in the form of modern technological inputs. These relatively high income farmers market greater shares of their product than their counterparts in other states. The data point to two anomalies which warrant clarification. In M.P., there is a very high concentration of land in middle and large size holdings, and in Kerala, a relatively high increase in tractor ownership occurred during the years 1966-1972. This indicates that neither marketing of product nor accumulation of capital *in isolation* are sufficient conditions for the existence of the capitalist mode, and that other characteristics of capitalism must also be present. Both Kerala and M.P. exhibited only one feature of capitalism, and this proved insufficient to promote the stimulation of the others.

Capitalism and Economic Growth

It is a basic Marxist hypothesis that the particular economic and social environment of capitalism is conducive to economic growth.[14] Capitalism stimulates technological change because capitalist owners of the means of production have a demand for innovative, more efficient methods of production, and they also have the ability to adopt innovative technology. An elaboration of this thesis is presented below.

The Demand for Technology

The demand for technological innovation is the result of both the capitalist's underlying profit motive as well as competitive necessity. According to Marxist theory, the sole driving force of the capitalist is the motivation to maximize profits by appropriating the maximum "surplus value." In Marx's words:

> The expansion of value becomes his subjective aim, and it is only in so far as the appropriation of ever more and more wealth in the abstract becomes the sole motive of his operations, that he functions as a capitalist.
> (Marx in Howard and King 1976: 92)

This motivation, coupled with the recognition that technological innovation increases productivity of inputs and hence the potential surplus to be appropriated, provides the incentive for the capitalist owner of the means of production to innovate.

However, the capitalist has a demand for innovative technology also as a result of the pressure exerted upon him by the highly competitive environment within which he operates. The adoption of technically innovative methods by each individual capitalist results in a decrease in costs of production due to increasing labor productivity, as well as a pressure to lower prices in order not to lose his share of the market. The result is a frantic competition among capitalists, which, according to Heertji, "expels the producers who hesitate to introduce new methods of production because they are not able to lower their prices" (Heertji in Howard and King, 1976: 223). Therefore, it is not only out of choice, but also out of necessity, that each individual capitalist innovates, as continuous innovation becomes a condition for his survival.

The Ability to Innovate
    In the capitalist mode of production, the profit motive and the general competitive environment create incentives for technological innovation. This growth promoting tendency of capitalism is further reinforced by the ability of owners of the means of production to adopt innovative technology. This is done by way of the "surplus value" which is transformed into money through the sale of produced goods. In turn, this money is used to purchase capital equipment which increases productivity of inputs and hence increases the "surplus value" that can be appropriated in each consecutive round. Since this "surplus value" is not consumed but invested, each consecutive round yields greater potential for capital accumulation.
    Production for the market also enables the adoption of innovative and growth promoting production processes. First, response to market conditions will stimulate production of cash crops in maximum quantities subject to cost constraints. Maximization can best be accomplished through efficient production methods, using modern technology. Second, production for the market entails cash payment for the product, hence increases the financial ability of farmers to adopt innovative technology.
    The labor market also plays a critical role in the capacity for innovation. The alienation of individuals from the means of production increases the supply of wage labor in the labor force. In addition, the loss of subsistence plots tends to push the peasant into the market for wage labor, thereby increasing the availability of labor for firms which require hired labor inputs.[15] Mandle (1979) claims that this increased supply of labor stimulates production by enabling already existing industries to expand and new ones to emerge. A free labor force implies the availability of workers to satisfy the expanded manpower requirements associated with increased production.

Economic Growth
    How does the capitalist's incentive and ability to innovate translate into economic growth? To answer this question, it is useful to return to Kuznets's definition of economic growth (presented in chapter one) according to which three equally important components interact to produce economic growth, which is defined as:

> The long term rise in capacity to supply
> increasingly diverse economic goods to its
> population, this growing capacity based on
> advancing technology and the institutional and
> ideological adjustments that it demands.
>
> (Kuznets 1973: 165)

Although Kuznets acknowledges that long term economic growth requires "institutional and ideological adjustments," he stops short of specifying these necessary changes in detail. He merely states that modern economic growth is not compatible with "family enterprise, illiteracy and slavery." The Marxian view, as expressed by Mandle (1979), goes beyond this, and specifically hypothesizes that it is the institutional and ideological environment provided by capitalism that is conducive to technological change. The motivation of capitalists to accumulate more capital for the purpose of purchasing more means of production stimulates the individual capitalist to innovate, allowing the economy to reproduce itself on an ever increasing scale. Thus, "expanded reproduction" is made feasible by increasing capital accumulation which, in turn, is made possible by new innovations. Both product and income grow at a pace far surpassing the rate of population growth, resulting in an increasing income per capita.

It is of interest that the Marxist debate on the Indian mode of production ignores the causal relationship between the capitalist mode of production and economic growth, thereby reflecting an oversight of Marx's writings on technology and the capitalist environment in which technological change thrives. With the possible exception of Omvedt's 1981 study,[16] capitalism is not viewed as a progressive force. In fact, a consensus was reached on the destructive role of capitalism in pre-Independence India on the grounds that she was incorporated into the world capitalism system by virtue of its status as a colony of a capitalist power. To Alavi (1975), this state of colonialism implied a retardation of growth in India due to the extraction of surplus by Great Britain. Banaji (1972) calls this colonial capitalism "backward capitalism" because of its role in the obstruction of local surplus reinvestment, a concept enthusiastically received by participants in the debate. Bagchi (1976) elaborates this concept by introducing the "de-industrialization" of India by the colonial power, and attributes to this process the retardation of growth.

Chattophaya (1972) is less categoric in his criticism of colonial capitalism, since he allots it a dual role and accedes that it preserved as well as destroyed industry. In fact, colonial capitalism may be backward insofar as penetration by an outside colonial capitalist power may strengthen the pre-bourgeois mode of production, thereby preventing or retarding the development of the capitalism that promotes growth.[17] However, this distinction was not drawn by Banaji or his colleagues.

The issue of the role played by capitalism in economic growth reflects a total neglect of Marx's writings on India in which he revealed his position on the progressive nature of capitalism. It may be that his studies on India were ignored because he described the mode of production in Hindustan in a manner which, according to Anderson (1974), was far from the existent reality.[18] In fact, Marx's asiatic mode of production is rarely mentioned in the debate. However wrong Marx might have been in describing Asian conditions, the exclusion of all his writings on British India from the debate does not seem warranted. Marx wrote that Indian society and economy were characterized by stagnation which could not change on its own, given that the social system had no internal source of transformation. Therefore, he argued, British intervention was beneficial to India insofar as it introduced capitalism to a region where it otherwise would not appear. This was part of the British "regenerative mission," which Marx expected would result in the creation of a capitalist and hence progressive environment in India:

> England has to fulfill a double mission in India: one destructive, the other regenerating-the annihilation of the old Asiatic society and the laying of material foundations of Western society in Asia.
>
> (Marx quoted in Kiernan 1967: 168)

In support of Marx, we would argue that an historical review of India shows the positive relationship between external penetrations and change. This was true of the Persian, Greek and later Mogul invasions, of the spread of the Maura, Kushan and Gupta empires, and of the French and Portuguese bases in the southern continent.

Despite the extensive research by Marx on British India, participants of the debate chose to ignore it probably because Marxists have difficulty giving British colonialism and therefore capitalism credit for little other than underdevelopment. Even outside the Indian debate, few Marxists chose to interpret Marx's writings on growth, capitalism and technology in a progressive fashion. Baran, Amin, Emmanual and Frank, all writing in the Marxist tradition, have insisted on the negative role of capitalism in the Third World. Warren (1973) first identified the positive relationship between capitalism, industrialization and increased income per capita in developing countries. Today, there seems little doubt that Warren's thesis is correct: there is ample evidence that many developing countries have experienced high rates of economic growth while exhibiting a capitalist mode of production. Hence, Marx's positive relationship between capitalism and economic growth seems to be corroborated by empirical evidence in the world.

Capitalism and Economic Growth in India

In regions where the capitalist environment prevails, technological change is stimulated and economic growth results. By analogy then, regions not characterized by this capitalist environment experience less technological change and therefore less economic growth, even though an appropriate new technology might be available. This Marxian hypothesis, that capitalist class relations promote the adoption of modern technology, is consistent with the findings from the ten Indian states. During the green revolution years, Punjab and Haryana distinguished themselves from the other agricultural states in the level and rate of change of their income per capita (discussed in chapter two). It was precisely these states that were identified in Tables 8.1 through 8.4 as exhibiting more characteristics of the capitalist mode relative to the remaining states. It is impossible to determine which of the four characteristics of the capitalist mode played the leading role in producing the capitalist environment and consequently enhancing economic growth. It is likely that they all interacted to different degrees in different regions. However, it is suggested that the generalized commodity production that occurred in Punjab and Haryana is both a consequence as well as motivation for technological change. As more and more farmers entered the sphere of capitalist farming in these states, they did so with the expectation of reaping the higher profits associated with the green revolution. In order to do so, farmers were compelled to adopt new technologies, which in turn enabled them to produce greater marketable quantities, accumulate more capital and increase further production.

In Punjab and Haryana, the increased capital accumulation manifested itself in increased mechanization which lead to economies of scale in the production of agricultural goods. First, the size of productive units increased, reflecting the increase in labor productivity and therefore the concentration of more capital in the hands of individual capitalist farmers.[19] Second, there was a tendency towards centralization, that is, the concentration of capital in the hands of a few capitalists who, through the adoption of innovative practices, survived the competitive struggle.[20] This is supported by empirical evidence of concentration of land in the hands of medium farmers in these two states. This twofold process resulting from the occurrence of technological change in a capitalist environment was described on a theoretical level by Marx:

> ...[centralization] is the concentration of already created capitals, the sublimation of their individual independence, the expropriation of capitalist by capitalist, the *transformation of many smaller into a few larger capitals.* (italics mine)
>
> (Marx in Howard and King 1976: 223)

Punjab and Haryana, then, experienced economic growth as described by Kuznets. However, when Kuznets stops short of describing the environment in which this best occurs, Marxist theory fills the void. Accordingly, growth in Punjab and Haryana took place in the environment most conducive to it, one in which labor was free and mobile, in which capitalists were motivated by profit and produced essentially for the market using hired labor and mechanical inputs. The social and economic environment which existed in the remaining eight states to a large degree lacked these characteristics, therefore lacked those mechanisms which stimulated technological innovation and hence economic growth.

The Transition Process and the Labor Force

The identification of the pre-capitalist mode of production or an analysis of the *transition* process from a non-capitalist to a capitalist mode are beyond the scope of this study. However, it is suggested that in areas where elements of capitalism *already* emerged, the transition process was accelerated and aided by some particular features of the labor force.

The relationship between landowners and those working their land (relations of production) in pre-Independence Indian agriculture was the focus of many studies by scholars from various backgrounds.[21] There is a consensus about the broad characteristics of the Indian agrarian scene, the salient features of which are widespread petty tenancy, sharecropping and exploitation by the landlords and moneylenders. Small cultivators survived on subsistence production. Regional variations depended on the prevalence of the zamindari or ryotwari systems of revenue collecting rarely hired for wages, but instead cultivated subsistence size plots for personal consumption, paid exorbitant rents to landholders (and to intermediaries in the zamindari regions), and were generally immobile with respect to landlord and region (Walinsky 1977).

Several events following Independence in 1947 stimulated the freeing of the rural labor force. With large regional variations, the Zamindari Abolition and Tenancy Laws of the 1950s resulted in the increase of landless labor, the increase of tenant owners and the increase of farming with hired labor. According to Omvedt (1981: A142), the purpose of these laws was not to give land to tenants but rather to the landless, yet the result was a great decrease in tenanted land as poor tenants were expelled from the land while large, more powerful tenants were given the land they already cultivated. The eviction of smaller and poorer tenants occurred because landholders, who were paid substantially for their lost land, began cultivating the land using hired agricultural labor. This process not only increased the landless labor force, but according to Omvedt, laid the groundwork for the transformation of the new owners into capitalists (Omvedt 1981: A142).

This process of expansion of the free agricultural labor force represents the beginning of a trend which was further reinforced by the green revolution in those areas where this package of technologies was most widespread, such as Punjab and Haryana, and scattered districts in other states. The green revolution, let us recall, was adopted more readily in regions where the capitalist environment provided farmers with greater motivation and ability to innovate. Once adopted, the green revolution affected the labor force by increasing the number of landless workers and by stimulating greater worker mobility, thereby freeing agricultural workers in both senses of the term. This occurred in two ways. First, the green revolution increased the profitability of farming and thereby stimulated increased numbers of owners to cultivate their own land. Since tenants are less prone to innovate on rented land, given the risks involved, their output was lower, and consequently rents charged had to be substantially lower than potential profits under self-cultivation. Hence, due to the profit motive, great numbers of tenants were evicted and became alienated from income-generating means of production.

Second, the peculiarities of the demand for agricultural wage labor stimulated regional mobility. As discussed in chapter two, the green revolution was characterized by seasonal labor abundance and shortage, depending on the timing of multiple cropping practices. This variability in the labor demand necessitated seasonal labor migrations.

depending on the timing of multiple cropping practices. This variability in
the labor demand necessitated seasonal labor migrations.

Thus, both a push and a pull force acted on the landless rural worker
to make him increasingly mobile. Although these forces may have been
present elsewhere, they were most accentuated in green revolution
regions. They further reinforced the freedom and alienation of the labor
force which already existed, although perhaps to a lesser degree.

Capitalism and Welfare

It has been established above that the high growth states, Punjab and
Haryana, experienced the greatest proliferation of capitalist agriculture
during 1960-1975. The question posed here is whether this economic
growth lead to an improvement in the welfare of the people in these states.
According to Marx, although capitalism is best able to produce economic
growth, it is by no means the system most conducive to the wellbeing of a
population. Marx and Engles, in the Communist Manifesto, contend that
the progress of industrialization does not entail an improvement in the
conditions of life of the workers (Freedman 1961: 17, 26).[22] Mandle
points out that, in fact,

> the thrust of Marx's critique of capitalism lies
> precisely in the contrast between the expansion of
> the capacity to produce and the limited extent to
> which those benefits will be realized by the
> population within that system.
>
> (Mandle 1982: 26)

A body of empirical literature has addressed the relationship between
capitalism and welfare in an effort to determine the relative welfare
indicators in different economic systems. Among these studies, Ahluwalia
(1974), Gurley (1979), Horvat (1974) and Mandle (1980) all agree that
socialist countries provide greater welfare to their populations.[23]
Ahluwalia presented empirical evidence to support his proposition that
socialist countries enjoy a greater degree of internal income equality.
Equality was also adopted as an indicator of welfare by Gurley, who
claimed that on the basis of it, as well as poverty and employment,
socialist countries performed slightly better than capitalist countries,
whereas on the basis of literacy and health, their performance by far
exceeded non-socialist countries. Horvat and Mandle both approach

conclude that socialist countries satisfied the basic needs of their populations more than capitalist ones.

The evidence from the agricultural states of India presented in chapter six shows that the relationship between growth and welfare is not necessarily direct. In fact, the high growth capitalist states of Punjab and Haryana did not distinguish themselves from the remaining eight states in the provision of social services to their people. In addition, although Kerala is a low growth state, it has by far the lowest infant mortality rate, as well as the highest number of hospital beds per population, highest literacy rates and among the highest degree of rural electrification.

The ten states of this study range over a wide spectrum of economic and political systems. As measured by the four indicators of capitalism, Punjab and Haryana ranked very high, whereas Kerala was at the opposite end of the spectrum, indicating a low proliferation of capitalist farming. However, it should be recalled from chapter six that this is the only state under study that had been governed by a communist government on various occasions since 1957. It was suggested that the provision of basic needs is more a consequence of policies regarding resource allocation than of growth. The evidence on basic needs and the degree of capitalism in agriculture points out the similarity of the Indian case with the international studies mentioned above. All in all, the Indian case seems to corroborate Marx's thesis that, although capitalism is the system most conducive to growth, it is by no means the one that best enhances the immediate welfare of the population.

Conclusion

The debate on the mode of production represents important exchanges among scholars trying to identify and formulate some pertinent questions about the Indian social and economic structure. The empirical evidence provided in this chapter adds to the debate insofar as it shows that the mode of production in the agriculture of Punjab and Haryana was unequivocally capitalist. While there is no doubt for those states, the mode of production in the other eight agricultural states remains questionable. We have not attempted to formulate an operational definition of the feudal, semi-feudal or colonial modes of production, hence an empirical assessment of those modes was not made. However, given the features of capitalism, the eight states exhibited some characteristics of capitalist agriculture, although not to the degree of Punjab and Haryana.

Notes

1. This ongoing debate began in earnest in 1969, although several authors from the early 1960s addressed issues relevant to the debate (for example, S.C. Gupta (1962), "Some Aspects of Indian Agriculture" in *Enquiry* 6, and D. Thorner (1967), "Capitalist Stirrings in Rural India" in *The Statesman*, Calcutta Nov. 1, 2, 3, 4).

2. The latter tended to focus more on capitalist global relations, whereas the Indian debate tended to limit itself to domestic matters.

3. Many students of Marxism have addressed issues pertaining to modes of production. See, among others, L. Althusser and E. Balibar (1970), *Reading 'Capital'*, London: New Left, and S. Clark et al. (1980), *One-Dimensional Marxism*, London: Allison and Busby.

4. In the ancient mode of production, labor is not free because of the prevalence of slavery. In the feudal mode, labor is not free because of a complex series of relationships which tie the peasant to a particular lord.

5. Rudra was the first to attempt a scientific precision uncharacteristic of Indian social science to date, and did so in reaction to studies which "consisted of no more than stray visits to the countryside and conversations with some farmers" (Rudra 1969: A143). He did, however, take it too far, in that he imposed such stringent conditions on the statistical correlations between variables that must exist prior to the qualification of a region as capitalist that it became impossible to satisfy them, so much so that the Punjab, in the midst of the green revolution, was viewed implausibly as non-capitalist.

6. See the extensive survey of the debate by Thorner 1982: 1964-1966.

7. The Indian debate represents the application to Indian conditions of some of the arguments arising from the debate between Frank and Laclau, which focused on South America and its transition from feudalism to capitalism (See D. Goodman and M. Redclift (1982), *From Peasant to Proletarian*, New York: St.Martin's Press, pp. 37-43). Frank entered the Indian debate briefly in 1973 only to criticize the participants for not considering the world economic system in their analysis of India ("On Feudal Modes, Models and Methods of Escaping Capitalist Reality" in *Economic and Political Weekly* 8/1).

8. In support of this argument, Patnaik (1971: A126) cites M.L. Darling as describing areas where the British government tried to "infuse

the capitalist element" and thereby strengthen the colonies by giving farmers grants to spend on improvement of their land. The result was, however, an unwillingness of the farmers to invest in the land, preferring instead consumption. In post-Independence India, farmers are more prone to invest in the land, hence the distinguishing feature of reinvestment of surplus into the sphere from which it was extracted.

9. Sau (1973) for instance, was not concerned with the sector in which capital accumulation takes place, so long as investment happens within national borders.

10. Among the other modes which have been identified in the Indian setting are the Asiatic (Marx), semi-feudal (Bhaduri, Prasad, Chandra, Sau and Rudra), colonial (Alavi) and dual (Lin). "Pre-capitalist" is also a term used often to connote any mode preceding capitalism.

11. Although it would have been desirable to sum the landless agricultural workers with those owning 2.5 or less acres and thereby comply with Omvedt's concept of wage labor, it was impossible due to data constraints. Plots by size were not divided by owner occupation, so it was impossible to distinguish between wage workers and cultivators.

12. This data pertains to the size of *operational* holdings, rather than *owned* holdings. Due to data unavailability, the former has been adopted as a proxy for the latter.

13. Gupta (1962) divided farmers into three categories, capitalist (operating farms over 20 acres), middle size market-oriented (operating farms between 10 and 20 acres) and small (operating farms less than 10 acres). He characterized the capitalist farmer as marketing the bulk of his product, and the middle size farmer of marketing at least 50 percent.

14. According to Mandle,

> In the Marxian paradigm it is the society's relations of production, broadly its class structure, which plays a causal role in the development of the productive forces; furthermore, it is argued that the class structure of some kinds of societies are more conducive to economic development than others.
>
> (Mandle 1982: 10)

Marx believed that the class structure of capitalism was most conducive to economic growth. Avineri describes Marx's view:

> ...what ultimately gave rise to the bourgeois
> world and industrialization was the slow
> differentiation of a sphere of activity which
> legitimized profit-oriented economic activity.
>
> (Avineri in Mandle 1982: 11)

15. Tenants may lose their plots as owners of that land opt for self-cultivation.

16. In her 1981 study, Omvedt describes the farmers in some parts of India as progressive insofar as they introduced new technology to production and thus contributed to the economic growth of the regions.

17. If penetration by a colonial power fails to transform any of the characteristics of the mode of production (for example, if production occurs with feudal serfs or slaves) then the movement towards capitalism, which Marx predicted would occur in all societies, would fail to proceed or even begin.

18. See P. Anderson (1974), *Lineages of the Absolutist State*, Atlantic Highlands: Humanities Press, for an excellent critique of the asiatic mode of production. Also, Marian Sawer (1977), *Marxism and the Question of the Asiatic Mode of Production*, The Hague: Martinus Nijhoff.

19. Omvedt (1981) describes the expansion of the land under cultivation by the big capitalist farmers as they lease land from small farmers who find it more profitable to rent out their land since they cannot afford the inputs necessary to compete in production. She refers to this trend as "reverse tenancy."

20. Dasgupta (1977: 249) claims: "Our study clearly shows that the skewness in the distribution of land is increasing [as a result of the green revolution]..." because big farmers increase land under cultivation so as to increase their profits.

21. Both Joshi and Ladejinsky have written extensively on the Indian agrarian structure. See, for example, P.C. Joshi (1979), "Agrarian Social Structures and Agrarian Reconstruction Patterns in Asia" in *Management of Agriculture: Collection of Theme Papers*, New Delhi: Indian Institute of Public Administration. Ladejinsky's work has been edited by L. Walinsky in 1977 (*Agrarian Reform as Unfinished Business*, London: Oxford University Press).

22. See Robert Freedman (ed.), (1961), *Marx on Economics*, New York: Harcourt, Brace and Co.

23. M. Ahluwalia (1974, "Income Inequality: Some Dimensions of a Problem" in *Finance and Development* (Sept.)), J. Gurley (1979,

"Economic Development: A Marxist View" in K.P. Jameson and C.K. Wilber (eds.), *Directions in Economic Development*, Notre Dame: University of Notre Dame) and B. Horvat (1974, "Welfare and the Common Man" in *World Development* 2/7)

9

Conclusion

This book has focused on the experience of the agricultural sector throughout the process of structural transformation of state economies. It has addressed many aspects of the *role* of agriculture, its *characteristics* and some of the *ramifications* of its growth. The linkages between these topics was provided by Kuznets's theory of modern economic growth, which described the main source of growth, the structural transformation that accompanies growth, and the environment in which this process occurs (chapter one). These linkages can now be amplified with empirical evidence from the agricultural states of India.

The principal source of growth in agriculture over the course of the past few decades was found to be technological innovation, as neither land nor labor increased in proportion to growth rates (chapter two). Punjab and Haryana adopted relatively more technology, a process that resulted in rates of growth greater than those experienced in other states. Various institutional factors were studied in order to determine why the high growth, wheat producing states were more innovative than the rice states. Even within the wheat regions, the experience of Punjab and Haryana differed from that of U.P. insofar as the farmers in the former states had greater access to capital and the prevailing size of operational holdings was appropriate for the green revolution inputs. It was thus clear from chapter two that the success of wheat producing states in the expansion of output and the increase in income was not merely the result of crop choice (i.e., wheat versus rice), but involved institutional factors as well. Further analysis of the innovations underlying agricultural growth in these states indicates that growth was by no means uniform: the prevailing technology was relatively more biochemical in Punjab, whereas in Haryana it was relatively more mechanical (chapter three).

The proliferation of innovative farming practices throughout rural India resulted in increases in income per capita in most regions. In chapter two, the interstate variation in growth rates was discussed, and the pace and pattern of income changes throughout agricultural India was identified. This growth entailed concomitant structural changes which altered the relative importance of agricultural income and labor in the state economies. Specifically, it was found that although all-India experienced a steady decrease in the proportion of income derived from agriculture over the past few decades, this was by no means true for the individual states. In other words, state level diffusion of agricultural growth into non-agricultural sectors did not take place in the same proportions as on the national level (chapter two). An inquiry into the industrial development in the most successful agricultural region, Punjab, revealed that the relative importance of industries did, in fact, increase following agricultural growth, and that those industries related to agriculture underwent extensive change during the past two decades (chapter five).

With respect to structural changes in the labor force, it was found that the short run effect of agricultural growth on the demand for agricultural labor was ambiguous, and depended upon the nature of the technological innovations (chapter three). In Punjab, where the biochemical tendencies were relatively more pronounced, an increase in the primary sector labor force was observed since the new technologies warranted a greater labor input for the performance of agricultural activities. In Haryana, where the mechanical tendencies of the green revolution were relatively stronger, the labor displacing effect of the technologies manifested itself in labor force changes by a decrease in the demand for agricultural workers. However, by the 1980s, this distinction between the two high growth states became less clear, coinciding with a convergence in the prevailing rural technologies. Furthermore, changes were observed in the composition of agricultural labor, such as an overall decrease in cultivators and an increase in agricultural laborers (chapter three).

Labor force changes with special reference to females were also identified. It was found in chapter four that female participation in the high growth agricultural regions was low compared to the national average, and decreased even further during the green revolution. This decrease occurred irrespective of the nature of the prevailing green revolution technology, whose effect was reflected in the nature of the work performed. The change in female participation in agriculture is explained by the interaction of economic growth and technological innovation as it occurred in the particular cultural setting of north India, indicating that innovation cannot be observed in isolation of the cultural context to which it is applied.

The discussion of structural changes in the labor force included a study of migration patterns which reflect interstate movement of population in response to labor market fluctuations. The evidence presented in chapter seven indicates the lack of a clear pattern of interstate migration: the principal losing and absorbing states are neither the high nor low growth agricultural regions. Instead, it was found that the industrial states of India tend to attract migrants from the agricultural states, and that interstate movements tend to occur in response to growth in the non-agricultural sectors, irrespective of their growth rates.

The last aspect of Kuznets's definition of modern economic growth relates to the environment within which growth occurs. Chapter eight contains an analysis of the capitalist mode of production in rural India, which was found to be most widespread in Punjab and Haryana. In addition, the economic and social environment concomitant with this mode of production was conducive to growth since it stimulated the supply and demand of technology. However, the capitalist mode of production did not provide the environment most conducive to improvements in human welfare, as evident from the empirical evidence of interstate variations in basic needs satisfaction and government expenditure on social services (chapter six).

An effort to tie together the findings presented above and to identify the interrelationships resulted in two major conclusions. The first pertains to the interstate differences in rates of development that clearly occurred throughout agricultural India, and the second is related to the growth promoting tendencies of capitalism. Both conclusions are discussed below, along with their policy implications. It is with great trepidation that policy recommendations are made since experience indicates that no single solution to problems associated with development is a panacea. The monumental effort, cost and organization that is necessary to ensure that real growth occurs usually outstrips the resources available to nations, forcing policy makers to construct a patchwork of second best solutions that are within the realm of possibility for developing economies.

Uneven Development

The empirical evidence presented throughout the book has confirmed that in the course of 1961 through 1981, the development experience across India has not been uniform. This was found to be true both among states as well as among individuals or groups within those states. As evidence of uneven regional development, the growth rates as well as the industrial distribution of income were compared, indicating that Punjab and Haryana were the fastest growing, yet that T.N. experienced the greatest decrease in the relative importance of agriculture coupled with an increase in industries. Furthermore, uneven development is evident by the interstate variations in human welfare improvements. The population of Kerala, followed by that of Punjab, seemed to enjoy the highest levels and improvements in living standards, while indicators of basic needs satisfaction showed little change in other agricultural states, *despite economic growth*. In addition, uneven development was evident in regional variations in technological innovation and the consequent labor force characteristics. Some states, namely Punjab and Kerala, experienced an increase in the primary sector labor force (both absolutely and proportionally) whereas other states experienced labor displacement which was usually not absorbed by industries and services. Clearly, the growth experience of the ten agricultural states differed greatly, in part as a result of different historical and institutional circumstances and in part because of the differing response to the green revolution technologies which enabled the gap between the states to expand.

What are some economic implications of this uneven development throughout agricultural India? What consequences can be envisioned for the individual states, be they the more or the less developed ones, and how does this uneven development affect the total functioning of the Indian economy? It is suggested that uneven growth, as reflected in regional variations in income and perhaps higher standards of living, is in and of itself not undesirable, and, as Hirshman proposed, may even be a condition for further *overall* economic growth (1958: 183). However, uneven development may become problematic when economic and political interaction *among states* and *with the center* occur under certain conditions. In an effort to identify some possible ramifications of uneven development, the economic interaction between agricultural states of varying development levels will be discussed. In addition, the economic and political implications of uneven development will be assessed.

Over time, regions develop by specialization in production according to comparative advantage, and thus a "division of labor" emerges among regions within a country, similar to what exists among nations. In India, the pattern that emerged *within the agricultural states* has been characterized by surplus production of foodgrains in Punjab and Haryana coupled with insufficient production in the remaining agricultural states (M.P. and Orissa have been surplus states only during several years between 1957 and 1980). One result of this pattern of regional development has been the export of surplus foodgrains from the highly productive regions in order to satisfy the demand in the deficit states. Can this situation continue *ad infinitum*? In response, two possible scenarios are suggested. The argument underlying both is that the tendency of uneven development among the two sets of states is self-perpetuating, even if the nature of their principal output changes over time.

(1) According to one scenario, the gap in foodgrain production among the deficit and surplus states will not close. As the deficit states increase their productivity, adopt new technology, and improve the output/cost ratio, the same will occur in the surplus states, although to a greater degree as a result of the advantageous conditions that these regions enjoyed in the first place. Assuming population growth will contribute to increasing demand for foodgrains equally in both deficit and surplus states, the gap in production will not diminish but in all likelihood will increase, thereby reinforcing the dependency of the deficit states. From the point of view of surplus states, uneven development is in their interest insofar as it guarantees markets for its foodgrains and thus stimulates the economy.

(2) The second scenario entails a decrease in demand from deficit states for agricultural product produced by surplus states. Such a saturation may have many sources, including the unlikely closing of the gap in agricultural production, or the substitution among consumer goods in the deficit states. The resulting decrease in dependency on the surplus state tends to stimulate agricultural production in the deficit states, at the expense of agricultural production in surplus states (assuming *ceteris paribus*). These surplus states, no longer faced with guaranteed markets, are compelled to compete in international markets and/or to switch into the production of other goods. If these new products are not agricultural, the process of substitution in production may stimulate the structural transformation of the economy to such a degree that the formerly surplus states undergo greater economic development than the deficit states and thereby perpetuate the existing gap between the two sets of states.

It thus seems that whichever of the two scenarios evolves, uneven development has a tendency to perpetuate itself in such a way that the more developed, or "leading" economies, increasingly differentiate themselves from the less developed ones with respect to growth and development indicators. Furthermore, the benefits of uneven development differ for the individual states and for the nation as a whole. The discussion of the scenarios served to show the different reactions high and low growth states might have to each other's production, omitting the role and reactions of the center as representative of national interests. What is the appropriate role for the central government to play when faced with uneven development among its states? Should federal policy encourage or discourage regional imbalances? An answer to these questions is based on a discussion of the economic and political effects of these imbalances.

Uneven regional development may have a positive effect on the national level if judged by purely *economic* implications of trade across state borders. This trade encompass Hirshman's (1958) trickle down effects or Myrdal's (1957) spread effects. The demand and supply in the labor and product markets are satisfied by the flow across regions: labor and raw materials flow into the high growth regions to satisfy input requirements of the production process, whereas output flows into the less developed regions to satisfy consumer and industrial demand. Without a difference in levels and nature of economic activity, these flows would not take place and requirements of the economy would have to be satisfied in international markets and/or at a higher cost.

From the perspective of *political economics,* uneven development may prove negative for the nation if (i) the more developed region attains economic self-sufficiency, or (ii) the central government attempts to redress the situation of inequality by providing advantages to the less developed regions, such as disproportionate tax burdens, disproportionate share of central funding, and price manipulations of goods. Either of these situations may cause the more developed region to reevaluate the relative costs and benefits of belonging to the national union. The costs may offset the benefits, and a decrease in economic ties with the national economy may result. At the same time, secession movements and other forms of animosity towards the center and the "nation" may emerge and upset center-state relations. These indications of dissatisfaction and potentially dramatic political change have negative repercussions not only on domestic economic events, but may also transcend national borders and affect the formulation of investment policy by international agents. Relative political stability is a prerequisite to foreign investment, and a high growth region that might otherwise provide a fertile investment opportunity, may prove to be a disastrous error of choice. Thus, uneven regional development may have far reaching effects on the national economy and polity.

The positive economic effects of uneven development may or may not be offset by the negative political effects, and the net result is determined by the relative strength and interaction of these two forces.

What was India's experience with respect to uneven development? The empirical evidence provided throughout this book supports the contention that the high growth regions of India have grown at a greater pace and thus increased the gap between themselves and the lower growth regions (chapter two). Their relationship with the remaining regions of India was characterized by the flow of factors of production (labor) as well as finished products (foodgrains). Government policy has aimed at redressing the unequal relationship between the surplus and deficit regions by installing price ceilings above which the product of the surplus states cannot be sold (chapter five). Lastly, in the most developed of the surplus regions, Punjab, resentment has been voiced concerning the perceived injustice underlying its economic relationship with the Indian union (chapter five).

Consequently, strategies that address uneven development must recognize the difficulties inherent in promoting agricultural development in some regions while redressing imbalances in others. Uneven growth need not be avoided, but careful consideration should be given to measures aimed at stimulating development in the lower growth regions, given their potential drawbacks for the economic union. It is these measures, which must differ according to the particulars of each situation, that can ensure the success or bring about the failure of development policy. Undoubtably, much research is still needed in order to formulate this policy with the necessary wisdom and comprehension of ramifications.

Growth and Human Welfare in Capitalism and Socialism

Based on empirical evidence from chapter eight, it is clear that an environment characterized by capitalist relations in production is conducive to economic growth but not to improvements in human welfare. Instead, economic systems most successful in attaining welfare goals are those where social issues are stressed. Therefore, conflict may arise between policies of growth promotion (most likely under capitalist conditions) and welfare enhancement (most likely in a socialist environment). In practice, the recognition of this conflict must underlie government policy. This entails the balancing of (i) the promotion of private enterprise, coupled with (ii) the simultaneous use of government revenue to provide programs for disadvantaged individuals.

The promotion of private enterprise consists of the creation of an environment in which the profit motive can be pursued, thus resulting in the proliferation of profitable technological advances. However, it is not only the proliferation of technology among producers that is important, but also the *creation,* or supply of this technology. In an environment characterized by profit maximization, demand for innovation stimulates its supply, *ceteris paribus.* This process has occurred throughout history. Currently, it can be observed in biotechnological research, in which the private sector plays the main role given that government resources are insufficient to take on a project of that magnitude. The creation of this new form of technology occurs in response to demand by farmers for new ways of increasing output, as well as the recognition of potential profits to be reaped by private firms. To foster a spirit of innovation and risk taking, governments must ensure a legal and economic climate receptive and accommodating of profit maximization.

It is precisely in this climate that government policy towards human welfare must coexist with private enterprise. Imbalances between individuals and groups within society should be rectified by encouraging the disadvantaged to compete without tampering excessively with the market mechanism.

A plan for the coexistence of capitalism and socialism is not offered here, but it is suggested that the recognition of the necessity to mix these two modes of production is a prerequisite to the simultaneous achievement of goals such as long term economic growth and human wellbeing. In this respect, Indian development policy approaches this coexistence insofar as private initiative is encouraged while social goals are pursued. However, Indian policy since Independence has been characterized by too much control and intervention in the private sector (as exemplified by the system of industrial liscencing), coupled with inappropriate and ineffectual interventions to aid the disadvantaged population. These measures have not been narrow in scope, but as a result of firmly entrenched power relationships, they have gone largely unnoticed outside of their origin in New Delhi (as in the case of land reforms). The result is that India has not achieved the rates of growth experienced by some capitalist developing countries, nor the success in social indicators of the socialist countries.

Clearly, a change in policy is warranted. This need was foremost in the mind of Rajiv Gandhi when he implemented changes such as the liberalization of the private sector. It is with great expectations that the economic and social results of this shift in policy are awaited.

List of Abbreviations

A.P.	Andhra Pradesh
CD	community development
CSO	Cental Statistical Organization
CPI	Communist Party of India
DIC	District Industries Center
FYP	Five Year Plan
HMT	Hindustani Machine Tools Co.
HYV	high yielding varieties
IADP	Intensive Agricultural Develpment Program
IRDP	Integrated Rural Development Program
M.P.	Madhya Pradesh
NIC	newly industrialized country
NSS	National Sample Survey
PAU	Punjab Agricultural University
RBI	Reserve Bank of India
SDP	state domestic product
T.N.	Tamil Nadu
U.P.	Uttar Pradesh

Bibliography

Primary Sources

Fertilizer Association of India, Statistics Division, *Fertilizer Statistics*, New Delhi, yearly volumes.

Government of India, Ministry of Agriculture and Irrigation, (1975), *Farm (Harvest) Prices of Principal Crops in India*, New Delhi.

Government of India, Ministry of Home Affairs, (1979), *Geographic Distribution of Internal Migration in India*, New Delhi: Office of Registrar General and Census Commissioner.

Government of India, Ministry of Home Affairs, (1961, 1971, 1981), *Indian Census*, New Delhi: Office of Registrar General and Census Commissioner.

Government of India, Ministry of Information and Broadcasting, *India*, New Delhi, yearly volumes.

Government of India, Ministry of Labor, *Indian Labour Yearbook*, New Delhi, yearly volumes.

---- (1978), *Rural Labor Enquiry*, New Delhi.

Government of India, Ministry of Planning, *Statistical Abstract*, New Delhi: Central Statistical Organization, yearly volumes.

Government of Punjab, Economic and Statistical Organization, *State Statistical Abstract*, Chandigarh, yearly volumes.

---- (1981), *Farm Accounts in Punjab, 1979-1980*, Chandigarh: Economic and Statistical Organization Publication #353.

Naidu, I.J. (1975), *All India Report on the Agricultural Census 1970-71*, New Delhi: Ministry of Agriculture and Irrigation.

Reserve Bank of India, (1981), "Analysis of Estimates of State Domestic Product" in *Reserve Bank of India Bulletin*, September.

---- (1978), "Analysis of Estimates of State Domestic Product" in *Reserve Bank of India Bulletin*, April.

World Bank, *World Development Report*, New York: Oxford University Press, yearly volumes.

Secondary Sources

Agarwal, Bina (1985), "Women and Technological Change in Agriculture: The Asian and African Experience" in I. Ahmed (ed.), *Technology and Rural Women*, London: George Allen and Unwin Pub.

---- (1983), *Mechanization in Indian Agriculture: An Analytical Study Based on the Punjab*, New Delhi: Allied Pub.

Ahmed, I. (ed.), (1985), *Technology and Rural Women*, London: George Allen and Unwin Pub.

Alavi, Hamza (1975), "India and the Colonial Mode of Production" in *Economic and Political Weekly* 10/33,34,35.

Ali Baig, M. (1978), "The Tractor in India" in *Farm Mechanics: Problems and Prospects*, New Delhi: Proceedings of Indian Society of Agricultural Engineers.

Avineri, Shlomo (1969), *Karl Marx on Colonialism and Modernization*, New York: Anchor Books.

Bagchi, A.K. (1982), *The Political Economy of Underdevelopment*, Cambridge: Cambridge University Press.

Bahduri, Amit (1973), "A Study in Agricultural Backwardness Under Semi-Feudalism" in *Economic Journal* 83.

Bairoch, Paul (1975), *The Economic Development of the Third World Since 1900*, Berkley: University of California Press.

Bal, H.S. (1974), "Impact of Mechanization on Farm Labour Employment" in *Agricultural Situation in India* 29.

Balasubramanyam, V.N. (1984), *The Economy of India*, London: Weidenfeld and Nicolson.

Banaji, Jairus (1972), "For a Theory of Colonial Modes of Production" in *Economic and Political Weekly* 7/52.

Bandyopadhyay, D. (1986), "Land Reforms in India" in *Economic and Political Weekly* 21/25-26.

Bardhan, P. (1984), *Land, Labor and Rural Poverty*, New York: Columbia University Press.

---- (1979), "On Class Relations in Indian Agriculture" in *Economic and Political Weekly* 14/19.

Bartsch, William (1977), *Employment and Technological Choice in Asian Agriculture*, New York: Praeger.

Becker, C., E. Mills and J.Williamson (1986), "Modeling Indian Migration and City Growth 1960-2000" in *Economic Development and Cultural Change* 35/1.

Beneria, Lourdes (ed.), (1982), *Women and Development: The Sexual Division of Labour in Rural Societies*, New York: Praeger.

Bhalla, Surjit (1979), "Farm and Technical Change in Indian Agriculture" in R. Berry and W. Cline (eds.), *Agrarian Structure and Productivity in Developing Countries*, Baltimore: Johns Hopkins University Press.

Bhat, M., S. Preston and T. Dyson (1984), *Vital Rates in India 1961-1981*, National Research Council, Washington: National Academy Press.

Bhattacharya, Sib Nath (1980), *Rural Industrialization in India*, New Delhi: B.R. Pub. Co.

Billings, Martin and Arjan Singh (1970), "Mechanization and the Wheat Revolution: Effects on Female Labor in the Punjab" in *Economic and Political Weekly* 5/50.

---- (1969), "Labour and the Green Revolution: The Experience in Punjab" in *Economic and Political Weekly* 4/52.

Binswanger, Hans (1978), *The Economics of Tractors in South Asia: An Analytical Review*, New York: Agricultural Development Council and the International Crops Research Institute for the Semi-arid Tropics.

Boserup, Esther (1970), *Women's Role in Economic Development*, London: Allen and Unwin.

Buttel, F., M. Kenney and J. Kloppenberg (1985), "From Green Revolution to Biorevolution: Some Observations on Changing Technological Bases of Economic Transformation in the Third World" in *Economic Development and Cultural Change* 34/1.

Chaduri, D.P. (1968), "Education and Agricultural Productivity in India," PhD. dissertation, University of Delhi.

Chandna, R.C. (1967), "Female Working Force in Punjab" in *Manpower Journal* 2/4.

Chattopadhyay, Paresh (1972), "On the Question of the Mode of Production in Indian Agriculture: A Preliminary Note" in *Economic and Political Weekly* 7/13.

Chawdhari, T.P.S. and B.M. Sharma (1961), "Female Labour of the Farm Family in Agriculture" in *Agricultural Situation in India* 6.

Chawla, J.S., S.S. Gill and R.P. Singh (1972), "Green Revolution, Mechanization and Rural Employment" in *Indian Journal of Agricultural Economics* 27/4.

D' Mello, L. (1979), "Agro-Based Industries: Feedback and Prospects-Cotton, Jute and Sugar" in C.H. Shah (ed.), *Agricultural Development of India, Policy and Problems*, New Delhi: Orient Longman.

Danatwala, M.L. (1979), "Agricultural Policy in India Since Independence" in C.H. Shah (ed.), *Agricultural Development of India, Policy and Problems*, New Delhi: Orient Longman.

Dasgupta, B. (1977), "India's Green Revolution" in *Economic and Political Weekly* 12/6,7,8.

Day, R.H. and Inderjit Singh (1977), *Economic Development as an Adaptive Process: The Green Revolution in the Indian Punjab*, New York: Cambridge University Press.

De Souza, Alfred (ed.), (1975), *Women in Contemporary India and South Asia*, New Delhi: Manohar.

Desai, Gunvant (1971), "Some Observations on Economics of Cultivating High-Yielding Varieties of Rice in India" in *Artha-Vikas* 7/2.

Desai, M., S.H. Rudolph and A. Rudra (eds.), (1984), *Agrarian Power and Agricultural Productivity in South Asia*, Berkeley: University of California Press.

Eicher, C. and Staatz J. (eds.), (1984), *Aricultural Development in the Third World*, Baltimore: Johns Hopkins University Press.

Feder, G., R. Just and D. Zilberman (1985), "Adoption of Agricultural Innovations in Developing Countries: A Survey" in *Economic Development and Cultural Change* 33/2.

Fic, Victor (1970), *Kerala, Yenan of India: Rise of Communist Power 1937-1969*, Bombay: Vora Pub.

Franda, Marcus (1979), *India's Rural Development*, Bloomington: Indiana University.Press.

Frankel, Francine (1971), *India's Green Revolution: Economic Gains and Political Costs*, Princeton: Princeton University Press.

Friedman, John (1972), "A Generalized Theory of Polarized Development" in N. M. Hansen (ed.), *Growth Centers in Regional Economic Development*, New York: The Free Press.

Ghosh, Bahnisikha and Sudhin Mukhopadhyay (1984), "Displacement of the Female in the Indian Labour Force" in *Economic and Political Weekly* 19/47.

Gibbons, David et al. (1980), *Agricultural Modernization, Poverty and Inequality*, Westmead: Teakfield Ltd.

Goetsch, Carl (1977), "Agricultural Mechanization in the Punjab: Some Comparative Observations from India and Pakistan" in R. E. Frykenberg (ed.), *Land, Tenure and Peasant in South Asia*, New Delhi: Orient Longman.

Gosal, G.S. and G. Krishan (1975), "Patterns of Internal Migration in India" in A. Leszek et al. (eds.), *People on the Move: Studies on Internal Migration*, London: Methuen and Co.

Griffin, K. (1974), *The Political Economy of Agrarian Change*, Cambridge: Harvard University Press.

Gulati, Leela (1984), "Technological Change and Women's Work Participation and Demoghraphic Behaviour" in *Economic and Political Weekly* 19/49.

---- (1975a), "Female Work Participation: a Study of Interstate Differences" in *Economic and Political Weekly* 10/2.

---- (1975b), "Occupational Distribution of Working Women" in *Economic and Political Weekly* 10/43.

Gupta, D.P. and K.K. Shangari (1980), *Agricultural Development in Punjab*, New Delhi: Agricole Publishing Academy.

Gupta, L.C. (1983), *Growth Theory and Strategy: New Direction*, New Delhi: Oxford University Press.

Heertije, Arnold (1976), "An Essay on Marxian Economics" in M.C. Howard and J.E. King (eds.), *The Economics of Marx*, Harmondsworth: Penguin Books.

Hicks, Norman (1979), "Growth vs Basic Needs: Is There a Trade-Off?" in *World Development* 7/11-12.

Hobsbawn, E.J. (ed.), (1975), *Karl Marx: Pre-Capitalist Economic Formations*, New York: International Publishers.

Holdcroft, Lane (1984), "The Rise and Fall of Community Development 1950-1965" in C. Eicher and J. Staatz (eds.), *Agricultural Development in the Third World*, Baltimore: Johns Hopkins University Press.

Howard, M.C. and J.E. King (ed.), (1976), *The Economics of Marx*, Harmondsworth: Penguin Books.

Jeffrey, R. (1986), *What's Happening to India? Punjab, Ethnic Conflict, Mrs. Gandhi's Death and the Test for Federalism*, London: Macmillan.

Johar, R.S. and J. S. Khanna (eds.), (1983), *Studies in Punjab Economy*, Amritsar: Guru Nanak Dev University.

Johar, R.S. and V. Chadha (1983), "Structure, Working and Performance of Public Industrial Enterprises in Punjab" in R.S. Johar and J.S. Khanna (eds.), *Studies in Punjab Economy*, Amritsar: Guru Nanak Dev University.

Johar, R.S. and P. Kumar (1983), "Imperative and Instruments of Industrialising Punjab" in R.S. Johar and P. Kumar (eds.), *Studies in Punjab Economy*, Amritsar: Guru Nanak Dev University.

Johl, S.S. (1973), "Mechanization, Labour Use and Productivity in Agriculture" in *Agricultural Situation in India* 28.

Johnston, B. and J. Cownie (1969), "The Seed and Fertilizer Revolution and the Labor Force Absorption Problem" in *American Economic Review* 59/4.

Joshi, P.C. (1975), *Land Reforms in India*, Bombay: Allied Pub. Ltd.

Kahlon, A.S. and D.S. Tyagi (1985), *Agricultural Price Policy in India*, New Delhi: Allied Pub. Ltd.

Khanna, G. and M. Varghese (1978), *Indian Women Today*, New Delhi: Vikas Publishing House Pvt. Ltd.

Khurso, A.M. (1973), *Economics of Land Reform and Farm Size in India*, New Delhi: Macmillan.

Kiernan, V.G. (1967), "Marx and India" in *The Socialist Register*, London: Merlin Press.

Krishna, Raj (1975), "Measurement of Direct and Indirect Employment Effects of Agricultural Growth with Technical Change" in L.G. Reynolds (ed.), *Agriculture in Development Theory*, New Haven: Yale University Press.

---- (1974), "Measurement of Direct and Indirect Employment Effects of Agricultural Growth with Technological Change" in E.D. Edwards (ed.), *Employment in Developing Nations*, New York: Columbia University Press.

---- (1967), "Agricultural Price Policy and Economic Development" in H.M. Southworth and B.F. Johnston (eds.), *Agricultural Development and Economic Development*, Ithaca: Cornell University Press.

Krishna, Raj and G.S. Ray Chaudhari (1982), "Trends in Rural Savings and Capital Formation in India, 1950-51 to 1973-74" in *Economic Development and Cultuiral Change* 30/2.

Krishnamurty, J. (1984), "Changes in the Indian Work Force" in *Economic and Political Weekly* 29/50.

Krishnan, T.N. (1977), "The Marketed Supply of Foodgrains" in P. Chandhuri (ed.), *Readings in Indian Agricultural Development*, London: Allen and Unwin.

Kuznets, Simon (1973), *Population, Capital and Growth*, New York: W.W. Norton and Co.

---- (1966), *Modern Economic Growth: Rate, Structure and Spread*, New Haven: Yale University Press.

Lee, Everett, Simon Kuznets and Hope Eldridge, et al., (1957, 1960, 1964), *Population Redistribution and Economic Growth, United States, 1870-1950*, vol. I, II, III, Philadelphia: American Philsosphical Society.

Lele, Uma (1986), "Women and Structural Transformation" in *Economic Development and Cultural Change* 3/2.

---- (1979), "Marketing and Pricing of Foodgrains" in C.H. Shah (ed.), *Agricultural Development of India*, New Delhi: Orient Longman.

Lin, Sharat G.(1980), "Theory of a Dual Mode of Production in Post-Colonial India" in *Economic and Political Weekly* 15/10,11.

Lipton, Micheal (1984), "Comment" in G. Meier and D. Seers (eds.), *Pioneers in Development*, Baltimore: Johns Hopkins University Press.

---- (1977), *Why Poor People Stay Poor*, Cambridge: Harvard University Press.

Lontfi, Martha Fetherolf (1983), *Rural Women: Unequal Partners in Development*, Geneva: International Labour Office.

Madhavan, M. C. (1985), "Indian Emigrants: Numbers, Characteristics Impact" in *Population and Development Review* 11/3.

Mahalanobis, P.C. (1953), "Science and National Planning" in *Sankya* 20/1-2.

Mandle, J.R. (1984a), "Carribean Dependency and Its Alternatives" in *Latin American Perspectives* 11/42.

---- (1984b), "Overcoming Dependency," unpublished paper.

---- (1982), *Patterns of Carribean Development*, New York: Gordon and Breach.

---- (1980), "Basic Needs and Economic Systems" in *Review of Social Economy* 38/2.

---- (1978), *The Roots of Black Poverty*, Durham: Duke University Press.

Marx, Karl and Frederick Engels (1970), *Selected Works*, Moscow: Progress Publishers.

Mascarenhas, R.C. (1982), *Technology Transfer and Development: India's Hindustan Machine Tools Company*, Boulder: Westview Press.

Mathews, R.G. (1984), "Initial Growth Pains in the Development of the Indian Machine Tool Industry" in *Indian Economic Journal* 32/2.

McEachern, D. (1976), "The Mode of Production in India" in *Journal of Contemporary Asia* 6/4.

Mellor, J. (1976), *The New Economics of Growth*, Ithaca: Cornell University Press.

---- (1969), "Agricultural Price Policy in the Context of Economic Development" in *American Journal of Agricultural Economics* 51/5.

Mellor, J. and G. Desai (eds.), (1986), *Agricultural Change and Rural Poverty*, New York: Oxford University Press.

Melotti, Umberto (1977), *Marx and the Third World*, London: Macmillan.

Mickelwait, Donald and Mary Riegelman (1976), *Women in Rural Development*, Boulder: Westview Press.

Mitra, A. (1968), "A Note on Internal Migration and Urbanization in India, 1961" in P. Sen Gupta and G. Sdasyuk (eds.), *Economic Regionalization of India: Problems and Approaches*, Office of Registrar General, New Delhi: Government of India Press.

Morawetz, David (1977), *Twenty Five Years of Economic Development 1950-1975*, Washington: World Bank.

Mukherjee, M. (1969), *National Income of India: Trends and Structure*, Calcutta: Statistical Publishing Society.

Narang, A.S. (1983), *Storm Over the Sutelj: The Akali Politics*, New Delhi: Gitanjali Publishing House.

Nath, Kamla (1970), "Female Work Participation and Economic Development" in *Economic and Political Weekly* 5/21.

Nayar, P.K.B. (1972), "Planning, Politics and People in Kerala" in P.K.B. Nayar (ed.), *Development of Kerala*, Trivandrum: University of Kerala.

Neale, W. (1985), "Indian Community Development, Local Government, Local Planning and Rural Policy Since 1950" in *Economic Development and Cultural Change* 33/4.

Nossiter, T.J. (1983), *Communism in Kerala: A Study in Political Adaptation*, Bombay: Oxford University Press.

Oberai, A.S. and H.K. Manmohan Singh (1980), "Migration Flows in Punjab's Green Revolution Belt" in *Economic and Political Weekly* 15/13.

Omvedt, Gail (1981), "Capitalist Agriculture and Rural Classes in India" in *Economic and Political Weekly* 16/52.

Pachal, T.K. (1980), "Internal Migration in India: Pattern, Implications and Policies" in *Demography India* 1-2.

Palmer, I. (1978), "Women and Green Revolutions," paper presented to the Conference on the Continuing Subordination of Women and the Development Process at the Institute of Development Studies, Sussex.

Parthasarathy, G. and D.S. Prasad (1978), *Response to the Impact of the New Rice Technology by Farm Size and Tenure: Andhra Pradesh, India*, Los Banos: International Rice Research Institute.

Patnaik, Utsa (1971), "Capitalist Development in Agriculture" in *Economic and Political Weekly* 6/39.

Prahaladachar, M. (1983), "Income Distribution Effects of the Green Revolution in India: A Review of the Empirical Evidence" in *World Development* 11.

Premi, Mahendra (1982), *The Demographic Situation in India*, Honolulu: East-West Center.

Randhawa, M.S. (1974), *Green Revolution: A Case Study of Punjab*, New Delhi: Vikas Pub. House Pvt. Ltd.

Rao, Amiya (1986), "Cost of Development Projects: Women Among the Willing Evacuees" in *Economic and Political Weekly* 21/12.

Rao, Hanumantha (1975), *Technological Change and Distribution of Gains in Indian Agriculture*, New Delhi: Macmillan.

Rao, R.S. (1970), "In Search of the Capitalist Farmer: A Comment" in *Economic and Political Weekly* 5/51.

Rao, V.K.R.V. (1983), *India's National Income, 1950-1980*, New Delhi: Sage.

Rao, V.M. and M. Vivekananda (1983), "Food Problem and Policy Priorities" in C.H. Shah (ed.), *Agricultural Development of India, Policy and Problems*, New Delhi: Orient Longman.

Reddy, N. (1975), "Female Work Participation: A Study in Interstate Differences-A Comment" in *Economic and Political Weekly* 10/23.

Rogers, Barbara (1980), *The Domestication of Women*, London: Tavistock Publications.

Rubin, B. (1985), "Economic Liberalization and the Indian State" in *Third World Quarterly* 7/4.

Rudra, A. (1981), "Against Feudalism" in *Economic and Political Weekly* 16/52.

---- (1970), "In Search of the Capitalist Farmer" in *Economic and Political Weekly* 5/26.

Sau, Ranjit (1973), "On the Essence and Manifestation of Capitalism in Indian Agriculture" in *Economic and Political Weekly* 8/13.

Schultz, T.W. (1981), *Investing in People: The Economics of Population Quality*, Berkeley: University of California Press.

Sen, Banhudas (1974), *The Green Revolution in India*, New York: Wiley Eastern Private Ltd.

Sen, Gupta P. (1968), "Some Characteristics of Internal Migration in India" in P. Sen and G. Sdasyuk (eds.), *Economic Regionalization of India: Problems and Approaches*, New Delhi: Office of Registrar General, Government of India Press.

Shackle, C. (1984), *The Sikhs* London: The Minority Rights Group Report No. 65.

Sharma, A.C. (1976), *Mechanization of Punjabi Agriculture*, New Delhi: Eurasia Publishing House.

Sharma, Ursula (1980), "Purdah and Public Space" in Alfred DeSouza (ed.), *Women in Contemporary India and South Asia*, New Delhi: Manohar.

---- (1978), "Women and Their Affines: The Viel as a Symbol of Seperation" in *Man* 13.

Sidhu, D.S. (1979), *Price Policy for Wheat in India*, New Delhi: S.Chand Co.

Sidhu, S. (1976), "The Structural Value of Education in Agricultural Development," Department of Agricultrual and Applied Economics Staff Paper 76, St.Paul: University of Minnesota.

Sims, H. (1986), "The State and Agricultural Productivity" in *Asian Survey* 26/4.

Singh, I. (1979), "Small Farmers and the Landless in South Asia," World Bank Staff Working Paper 320, Washington: World Bank.

Sinha, S.K. (1975), "Internal Migration (1971) and Popluation Redistribution" in R.B. Chari (ed.), *Demographic Trends in India*, New Delhi: Sunlight Pub.

Srinivasan, K. (1983), "India's Demographic Trends" paper presented at the Demography Association of India Conference, Bangalore, India.

Streeten, Paul (1979), "Basic Needs: Premises and Promises" in *Journal of Policy Modelling* 1/1.

Stromberg, Ann and Shirley Harkness (eds.), (1978), *Women Working: Theories and Facts in Perspective*, New York: Mayfield Pub. Co.

Sukumaran Nayar, V.K.S. (ed.), (1969), *Kerala Society and Politics* Trivandrum: Indian Political Science Conference.

Thapar, S.D. (1972), "Industrial Employment in Punjab and Haryana: 1966-69," US AID Economic Affairs Division Staff Paper, New Delhi.

Thorner, Alice (1982), "Semi-Feudalism or Capitalism- Contemporary Debate on Classes and Modes of Production in India" in *Economic and Political Weekly* 27/49,50,51.

Tinker, Irene and Michele Bo Bramsen (eds.), (1976), *Women and World Development*,Washington: Overseas Development Council.

Uppal, J.S. (1983), "Agrarian Structure and Land Reform in India" in J.S. Uppal (ed.), *India's Economic Problems, An Analytical Approach*, New Delhi: Tata McGraw Hill.

Vaidyanathan, K. E. (1967), "Population Redistribution and Economic Change 1951-1961," PhD thesis, University of Pennsylvania.

Vyas, V.S. (1975), "India's High Yielding Varieties Program in Wheat, 1966/7 to 1971/2," working paper, Mexico City: Centro Internacional de Mejoramiento de Maiz y Trigo.

Warren, Bill (1973), "Imperialism and Capitalist Industrialization" in *New Left Review* 81.

Yesudas, R.N. (1980), "Christian Missionaries and Social Awakening in Kerala" in *Journal of Kerala Studies* 7/1-4.

Youssef, Nadia Haggag (1974), *Women and Work in Developing Societies*, Westport: Greenwood Press.

Yudelman, M. et al. (1971), *Technological Change in Agriculture and Employment in Developing Countries*, Paris: Organisation of Economic Co-operation and Development.

Zachariah, K. C. (1964), *A Historical Study of Internal Migration in the Indian Subcontinent 1901-1931*, Bombay: Asia Publishing House.

Zarkovic, M. (1987a), "Linkages With the Global Economy: The Case of Indian and Yugoslav Foodgrain Production," unpublished paper.

---- (1987b), "The Effects of Economic Growth and Technological Innovation on the Agricultural Labor Force in India" in *Studies in Comparative International Development* 22/4.

---- (1986), "Economic Growth and Basic Needs in India" in *Journal of South Asian and Middle Eastern Studies* 9/3.

---- (1984), *Economic Growth, Migration and Welfare in India's Agricultural States*, PhD thesis, Temple University.